Early Praise for *Become an Effective Software Engineering Manager*

James's book will be at the top of my mind when I start a product company. The content is not just for software engineering managers—technical leaders, CTOs, and managers will all benefit from practical advice that helps them be successful with the people they manage and work with.

➤ **Stephen Bussey**
Author of *Real-Time Phoenix*, Software Architect, SalesLoft

I am really grateful that finally a book exists on how to be a good technical manager, with engineering-level clear instructions. This book is the only reason I have survived my first two months on the job. I followed its advice on hiring, running one-to-ones, and most importantly, managing my own time. We hired some amazing people, and the hires and my sanity are still here—so this book must be teaching me something right.

➤ **Lev Konstantinovskiy**
Data Science Team Lead, Trading Firm

This is a book I wish every engineering manager read before they stepped into the role. Bite-sized pieces of advice on engineering management that you would expect from a Pragmatic Publishers book.

➤ **Patrick Kua**
Founder, Patkua.com

There still are very few books written by hands-on engineering managers working at tech companies on how to succeed in this kind of environment. This book is exactly that. The stories are relatable, the takeaways well articulated, and the exercises will help readers become better managers. It's the book I'll be recommending to new managers transitioning into the role, managers starting at a new company, and experienced managers looking to make an organization-wide impact.

➤ **Gergely Orosz**
Engineering Manager, Uber

James comes across as a sincere author who writes from his experience in management, and it shows in his writing.

➤ **Chris Dudley**
Engineer, Brandwatch

Become an Effective Software Engineering Manager

How to Be the Leader Your Development Team Needs

James Stanier

The Pragmatic Bookshelf
Raleigh, North Carolina

For our complete catalog of hands-on, practical, and Pragmatic content for software developers, please visit *https://pragprog.com.*

The team that produced this book includes:

Publisher: Andy Hunt
VP of Operations: Janet Furlow
Executive Editor: Dave Rankin
Development Editor: Adaobi Obi Tulton
Copy Editor: L. Sakhi MacMillan
Indexing: Potomac Indexing, LLC
Layout: Gilson Graphics

For sales, volume licensing, and support, please contact *support@pragprog.com.*

For international rights, please contact *rights@pragprog.com.*

ISBN-13: 978-1-68050-724-9
Book version: P2.0—March 2021

Contents

Part III — The Bigger Picture

Acknowledgments

Anything of worth is rarely achieved alone. The main actor is always surrounded by a diligent and capable supporting cast, able to make the performance that much better.

I'd like to thank everyone that I have worked with at Brandwatch over the last eight years. You are all incredibly smart and ambitious people that have transformed a little technology company in a quaint seaside town into a global success. I'm fortunate that I get to work with all of you every single day. I always learn something new and see different perspectives other than my own. After all, you all encouraged me to start writing a modest blog on this subject in the first place, and look where that's gotten us.

A special thanks to those that have managed me over the years—Fabrice Retkowsky, Dave Jones, and Chris Bingham—for giving me the opportunity to become a manager and then to keep growing, growing, and growing. Additional thanks must be given to everybody that I've been lucky enough to have report to me. That's now too many to list! You let me practice being a manager for real, enabling me to become better at all of the skills that I've written about in this book, to the point that I can share them with the wider world.

Naomi Trickey must be thanked for graciously letting me write about the competency tracks and the accompanying career development exercises that she created collaboratively with Amy Collins. I hope that I've done them justice and that others around the world will now see their huge benefit. Thank you to Ollie Glass for introducing me to Lev Vygotsky and his theories around learning.

My publisher, The Pragmatic Bookshelf, has been fantastic to work with. From the pitch through to the final product, I have felt supported and inspired to do more. Adaobi Obi Tulton, my development editor, has been instrumental at getting to the heart of the narrative and keeping me on the straight and narrow. I couldn't have done it without her.

A number of readers have given me essential feedback while I was writing this book. Thank you to Burhan Ali, Chris Dudley, David Jones, James Da Costa, Lev Konstantinovskiy, Steve Hunt, Jessica Burns, Stephen Bussey, Yanick Nedderhoff, Chris Johnson, Naomi Trickey, Patrick Kua, Silvia Alvaro, and Rafael Barbosa. Additionally, everyone that has read and commented on my articles over the years has made me a better writer.

Last, but not least, I'd like to thank Rebecca Harrison for, well, pretty much everything. You are my partner, my best friend, my confidant, and my catalyst for betterment in all aspects of my life. It just so happens you've done an incredible job of illustrating this book also.

Oh, Finch—our Welsh Terrier—deserves a mention as well, mostly because she'll know if I don't, despite the fact that she can't read. A stolen sock here and the occasional chaos there remind me that there's no need to be so serious about everything. Life can be a game if we choose it to be. You'll see her popping up from time to time.

And I almost forgot: thank *you* for picking up this book. I hope that it opens doors.

Introduction

So, you're here. Why is that? Think about it. Perhaps you're interested in becoming a manager in the future. Maybe you've been promoted into a new role and are looking for advice on how to get started. You may have already been a manager for some time and are looking for a guide to help you become better at your job. No matter where you are in your career, this book is for you. Yes, *you*.

The technology industry is facing a skills crisis. This isn't because we don't know how to write software or how to scale our infrastructure. We've been getting a lot better at that. Instead, it's because we don't know how to manage people. Computers don't create software, people do. We need to make more *people* succeed. Good managers can solve this problem. *You* can be a great one.

This book intends to help address the skills crisis by giving you the hands-on, practical management advice that you need. There's no business fluff or leadership bravado here. You'll find no anecdotes about the military, no heroics, no cheesy metaphors, and no talk of Stanford. This book is about the tools that you need—the actual skills to do the job. In the same way that you would turn to a book to learn a new programming language, this book is here to help you learn how to be a world-class manager.

The Righteous Path

The problem with management, especially in the technology industry, is that many of us that become managers haven't been planning to do so for our whole lives. The same can't be said about creating software. Many of us tinkered with building websites, did online programming tutorials, or sketched out pictures of what we thought great software would look like way before we did formal education or started getting paid for it.

Often, we don't learn about management until we enter industry and see other people doing it. We might not learn about it until we see that there's an opening to run one of the teams at our company. Regrettably, sometimes that choice is forced upon us as the only way to progress our careers.

This means that we haven't had enough preparation, we haven't done any formal education, and we haven't much of an idea about what we should be doing as managers. We can only learn on the job and from those around us, and we can't offer ourselves any guarantee that the way that we are managing others is the *right* way.

I certainly didn't plan to be a manager. When I joined a local startup, I had just finished my PhD in compilers. For the previous twelve months, I was trying to find an academic role, but I failed. I had to do something else. I decided to give writing back-end code a try instead. It was fun and I built some cool stuff. But I had no idea where I wanted my career to go. Would I be writing code forever until I retired? Was that my path?

As that startup grew rapidly, teams began to form. I asked whether I could be considered to manage one of those teams—mostly out of curiosity—and the rest is history.

I didn't know how to be a manager, so I bought books. Lots and lots of books. They were stacked on my desk in the office and also at home. Some were good, but many were bad. Few gave practical advice that I could actually implement in my job. It's all well and good learning how the CEO of a Fortune 500 company spends their day, but how did that apply to me? I just wanted to know I was doing the right thing for me and my team.

Searching online for advice was even worse: a mixture of contradictory information. Some of it was written by people clearly pushing their coaching services. Some felt wrong, old fashioned, or irrelevant. I didn't need *7 Ways to Motivate My Staff* nor did I need to know *The One Thing to Make Your Team a Success*. What I really needed to know was what I should be doing each day, week, and month that would make me a better manager and make my team better as a result.

That was almost nine years ago. Unfortunately, the situation hasn't changed much for new managers. I still talk to people who are running a team for the first time who find themselves without support, without suitable role models, and without the ability to confidently say that they are doing a good job. This is a tremendous shame, since managing people can be one of the most rewarding jobs out there.

Consider this: *you* can create the conditions that allow others to succeed, to learn, to feel psychologically safe, and to be creative. *You* can be the person who helps your staff grow and achieve way beyond the level they thought was possible. *You* can create the team that allows people to enjoy turning up to work each day, ready to tackle challenging problems together. Twenty years from now, *you* can be front of mind when your staff are asked where their careers began to take off. Yes, that really can be *you*. This book will show you how.

Management and Leadership

The word *management* has a bad rap. Management may conjure up images of stuffy people in suits carrying stacks of papers in leather binders. Layers of bureaucracy. Pointless activities to maintain high-paying jobs in a hierarchy. *Leadership*, however, may be the look you're striving for. Being the inspirational figure that others look up to. Giving a talk to your department to rapturous applause. Being on the 40 Under 40 list. However, these two words are not disparate. They are strongly linked.

Management is the method and tools that you use to perform your job as a manager. However, using them well, alongside being a humble and caring human being, will elevate you higher. You will act with respect, grace, and kind consideration and will provide honest, candid conversation. You'll practice what you preach through your actions, and *that* will make you into a leader.

That's because doing management diligently, skillfully, and empathetically is *hard*. It is a rare skill. But it's a skill that you can harness if you read these pages, learn some new concepts, and then go out and *do* them. Management is a craft, much like writing software. You'll get better every day. It's an art, not a science. It's a mixture of your creativity, your personality, your heart, your mind, your ethics, and your values. It is *you*. You can be a manager and leader that others look up to. That others want to work for. That you enjoy being every day.

Your Journey

You may be at different stages on your journey. You may have picked up this book because you were interested in whether being a manager is something that you might like to do. You may have just been promoted within your existing company. You may have taken a leap into being a manager somewhere completely new. It doesn't matter where you are: you will find value in these pages. Even if you never become a manager, you will understand better about how they work. You can empathize. Bond. We will discover plenty of tools and techniques that are applicable to *anyone*, regardless of what job they are doing.

However, for the purpose of this book, we're going to imagine that you've turned up for your first day of your first managerial role in a new company. We're going to start afresh together and build from the ground up. It's going to be fun, and I'm honored that we're going to go on that journey together.

This book is split into three parts. It begins prescriptively: you'll be asked to *try this* or *do that*. Give it a go. As the book progresses, we'll consider more abstract and nuanced issues of management of which there is no right answer. There's only you and your way of doing it. How you act in those situations is where you begin to understand the most about yourself in this role.

You don't have to read the book from the beginning to the end, although you're welcome to do that. Care has been taken to write this book in such a way that will allow you to read chapters independently when you need some guidance on that subject.

The Outline of This Book

In the first part of the book, *Getting Oriented*, you're going to drop right into your new job. In *A New Adventure* you'll be getting to grips with your new role and working out who your team are, what they do, and how they relate to the rest of the organization. Then you're going to learn how to *Manage Yourself First* by getting your routine and habits in order.

Then we're going to move on to the second part of the book, *Working with Individuals*. Here you're going to learn all of the necessary tools and processes to be a success in your day-to-day role. We'll look at the ways in which you will be *Interfacing with Humans* every day, focusing on how you can make those interactions fruitful and positive. You'll then learn how to begin doing weekly *One-to-Ones*, which is the bedrock of your relationship with your staff.

You'll also learn how we're all motivated differently in *The Right Job for the Person*, understanding how you can use this to help people succeed and be happy. Then, we'll get to the nitty-gritty. We'll look into hiring in *Join Us!* and people leaving in *Game Over*. We'll close out the second part of the book by considering how you can begin to increase your impact outside of your team in *How to Win Friends and Influence People*, where you'll find a number of ways that you can contribute to your department for the greater good.

The third part of the book is called *The Bigger Picture*. This is where you'll begin to experience the messier sides of management. We'll look at why *Humans Are Hard* and *Projects Are Hard* and what you can do in various tricky scenarios. Then we'll consider the information that you'll have to handle, share, and store in *The Information Stock Exchange*. We'll then dive deeper into the often-paradoxical psychology of being an effective manager in *Letting Go of Control.*

We'll then turn our focus to your environment. You'll learn ways in which you can make your department's communication more effective in *Good Housekeeping.* Then, we'll consider how to create career progression tracks for your staff in *Dual Ladders. The Modern Workplace* addresses inclusivity, diversity, and culture and what you can do to make your company a better place for everyone.

We'll finish off by considering *your* future. After all, this is a book about *you.* In *Startups* we'll explore whether you might want to try and accelerate your career by joining a small but fast-growing organization. Then, in *The Crystal Ball,* we'll run through an exercise together that you can use to think deeply about where your career is going. And guess what? You can use it with your own staff as well.

So, What's Next?

All that's left is for me to thank you for embarking on this journey together with me. I hope that you find it entertaining and informative and that it helps you become a better manager and maybe even a better person. All I know is that if I had this book when I started, I would have felt more confident, more comfortable, and more able to know that I was doing a good job for myself, but especially for those in my team. I truly hope that it does the same for you.

Let's get started then. Are you ready for your first day?

Part I

Getting Oriented

So, you're new here. That's no problem at all.

We're going to take you through what you should do during your first days in your new role: what to do, who to talk to, and what you need to find out. This will set you up for success.

Then, before we start managing others, we're going to look at managing yourself. You'll work on getting your house in order and installing a framework to enable you to be efficient and effective.

Right, let's get oriented!

If it scares you, it might be a good thing to try.

> Seth Godin

CHAPTER 1

A New Adventure

You hold the ID card out in front of you so you can read it. Fortunately they've spelled your name correctly. Shame about the photograph, though. You return it to your pocket. You've arrived. It's day one.

"I'm so glad you're here!" says an approaching figure with her arm extended to shake your hand. She does so vigorously. It's Lisa, your new boss.

Following your manager across the open-plan office, you look around and take in the sights and sounds of your new workplace. As you pass the kitchen, you hear the frothing of the milk wand on the espresso machine and the chatter of your new colleagues. A ping-pong ball ricochets off of the glass dividing wall. A man whizzes past on a scooter. This is quite different from your old workplace.

A whiteboard in the middle of the office is surrounded by a red velvet rope. Lisa turns around to face it. You look puzzled. It's crammed full of impenetrable mathematics. The bottom right contains a collection of signatures. Lisa points at it.

"This is the algorithm that gave birth to this company. The signatures are from our founders. Who'd have thought that such a beautiful series of calculations could create a unicorn?"

You nod and smile in acknowledgment, pretending to understand what it means.

Lisa continues walking toward the side of the office that has floor-to-ceiling windows. You can see her making a beeline toward an empty desk. She gestures toward your chair.

"Have a seat," she says. "I've got to run to a meeting, but send me a DM if you need anything. I'll pop over again later."

"Sure, thanks!" you reply. Lisa dashes off behind one of the whiteboards. Unlike the one in the center of the office, it's not covered in complex mathematical symbols. Instead, it has an incredibly bad drawing of a horse with some text underneath. You squint to read it.

"Hello new *neigh*-bor!"

Oh dear. You sit down. Your chair emits an almighty squeak. You move your arm. Another giant squeak. Well, that's going to get old fast.

Hitting the spacebar, your computer comes to life. You type your username and temporary password into the login prompt and hit return. You have two unread emails. Number one:

Hey,

I forgot to tell you my DM handle. It's @lisag. Shout if you need anything. And I'm sorry about that horse drawing. Glad to have you on board!

Lisa

The next message appears to be automated.

Welcome!

We are proud to have you as part of our company. Please click the link below in order to check your personal details are correct.

Click here *to check your details.*

An automated message from PeopleWare HR Solutions. Do not reply.

You click the link and are taken to a page to review your profile. Your eyes are drawn to the box denoting your role.

```
Job title: Engineering Manager
Team: Infrastructure
```

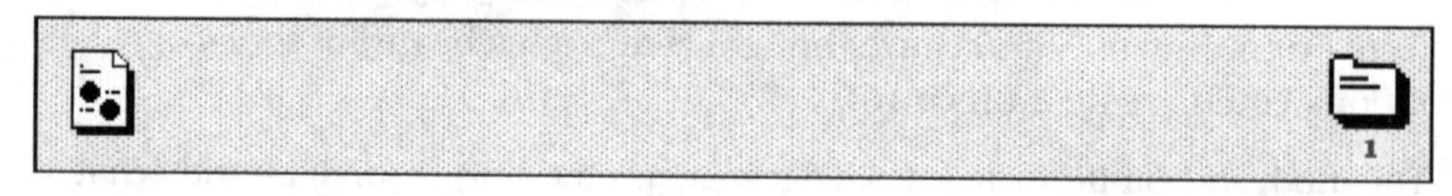

Hello, and welcome to your new job as a manager. It's great to have you here. We're going to get straight to business. It's your first week on your new job. How should you get started? What should you be doing? Help is at hand. We're going to give you some structure to make your first week a breeze.

Being a manager is a difficult job and those that become managers often have little practical guidance. Where's the API documentation for humans? Many engineers enter their first job having studied or practiced their skills for many years, either via education or self-directed learning. However, most new managers have never managed anyone before and often find themselves overwhelmed and unsure of how to do their job. This can be immensely stressful. I wonder how many potentially fantastic managers never continued down that path because they were unable to get the support that they needed. No more, I say. No more.

That's why this book is here. Throughout your journey, you're going to learn all of the essential skills you'll need to be the manager that you wish *you* had. This book is designed to incrementally teach you the practical, hands-on skills that you will need day-to-day, through to bigger-picture topics such as pressure, diversity, and career development.

We'll begin by getting you oriented during your first week as a manager, showing you how to get to know your team and your workplace, and give you the tools you need to manage yourself. Then we'll show you how to communicate effectively, how to do one-to-one meetings, consider how different people are motivated, then learn how to run performance reviews, hiring processes, letting people go, and consider how you can build visibility and influence.

Later, we'll consider stress, workplace politics, your own mental health, departmental best practice, career development, diversity, and inclusion. We'll conclude by looking at startups and planning out the future that you wish to have in your career.

However, let's get back to your first week. You're going to:

- Meet your team and your manager.
- Book in the meetings that you need.
- Create your snapshot of the team.
- Create an action list of items to focus on next.

Are you ready? Here we go!

Beginning Your First Week

Congratulations on your new role! You may have not made the transition into management in real life yet, but in this book you have, so well done. You are the star of this story.

Do you remember your first day in your current job? Perhaps you turned up at the office to discover that everything that you needed to know was documented in easy-to-find places, that the organization chart was up to date, that there was an informative onboarding process, and that building your development environment was an absolute breeze. Oh, wait. No, this is the real world. Instead, regardless of whether you joined the company to write Java, do QA, or to be a Scrum master, it's likely that you spent your first week muddling through, feeling like an impostor, and worrying whether you're making a good first impression with all of your new colleagues. Don't worry. It happens to us all.

Joining a new organization as a manager, or even being promoted from within your existing company, can be daunting. Joining as an engineer is equally scary, but at least you can ask for some bugs to fix while working out how the development environment is put together. You can find *something* productive to do straight away. When you join as a manager, what should you actually do? The initial feeling of pride that you experienced when accepting the job offer may quickly fade and be replaced with a bevy of questions:

- How should I get off on the right foot with my team?
- How is my relationship with others going to change now that I'm the boss?
- Why do I keep forgetting my pass when I visit the bathroom?
- What does the company expect me to do with my time?
- How am I meant to feel productive?
- What is my output?
- What am I actually responsible for?
- How is my performance measured?
- How on earth can I stop this chair squeaking?

Impostor Syndrome

Often, high-achieving individuals may experience an internal contradiction about their ability, resulting in a feeling like they are a fraud or a failure. This phenomenon is known as impostor syndrome. Individuals feel like they are faking it and that they will soon get found out.

If you feel this way as a new manager, then don't worry. It's normal. Know that your caution is because you want to do a good job for the benefit of others. You were given this role in the first place because you were qualified for it. Remember that. You will feel more confident with time. We'll look at impostor syndrome and its effects in *Humans Are Hard*.

Depending on the amount of preparation that you've had, you may have already come up with a plan for how you're going to spend your time. But for many new managers, it's common to feel like you're just making your job up as you go along, and more importantly, that you're somehow not qualified to do it in the first place. You may have had a number of managers over the course of your career that each demonstrated different styles and approaches: maybe you had a quiet, sage manager who was better at structured written communication when compared to face-to-face exchanges; maybe you had a bouncing entrepreneurial manager who motivated valiantly but was scattershot in their support, or maybe you had a manager who was consistently never there at all (believe me, I've heard stories). What should you do?

Don't worry. This book is here to help you. We'll begin by bringing some structure to your first week. Everything that you do in this chapter will feed into the activities that you'll learn about and master as the book progresses, but for now, just follow along with an open mind.

Creating a Snapshot

If you're coming from a programming background, then you may be familiar with the concept of *snapshots*: they're the state of a system at a given time. They are often used to allow a program to quickly restore from backup.

Since you're a management application that has just been deployed into production, you're going to need to restore from backup. But—oh no!—there isn't a backup to restore from. So you're going to need to create one. You're going to create a snapshot for yourself. This snapshot will be the basis on which you will begin your work with the team.

With time, you'll be able to form your own independent view of your team. You'll know how they are performing, the direction of their work in the future, and the individuals themselves. However, since you've only just started, you're going to have to do some investigative work. At any given time, there are three views of the team that you can compare and contrast. They can be used to discover areas you can work on right away.

Consider the following three areas.

These are:

- *Your own observations.* What you see day in, day out when working with your team. However, you've only just started, so you'll need to rely on others while you form your own view.
- *Your manager's observations.* Your manager will be interested in your performance, which fundamentally is your team's performance. What do they think about them and why?
- *Your team's observations.* This actually consists of the individual views of each member of the team, but we'll group them into one for simplicity. How do they feel about themselves and their colleagues? How do they feel as part of the wider company?

Your initial task is to work out what those views are. Once you've done so, then you can compare them against each other to discover where your work can begin.

Ready? Excellent. Let's get going.

Introduce Yourself to Your Team

First things first: it's time to say hello. Your team are the people through which you do your work as a manager. The relationship that you form with them is the foundation on top of which everything else is built. You'll want to meet each of them in the most natural way possible—in person or via a video call—rather than relying on email or chat software. You'll also want to take the time to meet them all one-to-one, rather than just saying hello in a group setting. The reason is that now you're going to begin *gathering information* from all around you, and you'll want to create a safe, private environment for each of your team to share their thoughts openly and honestly.

For each of your staff, ask them for thirty minutes of their time at their convenience this week. Whether you want to meet with them in a private room or go and grab a coffee is up to you: only you know what feels best in the context of your workplace and your team. However, you'll want to keep the meeting informal, since the two main aspects that you want to uncover will emerge best as part of an open conversation:

1. You'll want them to describe what the team is responsible for and what they are currently working on.
2. You'll also want to invite their opinion about what they like about the team and what is going well compared to what they currently find most challenging and needs improving.

These are broad questions, but given that you've only just started and that you're their new manager, you'll begin to uncover all sorts of information about your team.

For example, you will likely find out:

- Details of the current project(s) and how clear their objectives are.
- Knowledge, or confusion (!) about who is accountable for particular decisions, such as the priority of work and the roadmap.
- How they feel about their job.
- How they feel that their team is perceived within the department and the company.
- Their enthusiasm for their current project, and what they like and dislike about it.

You might even find out the secret to fixing your squeaky chair.

Breaking the Ice

One of the hardest things that I had to deal with when transitioning to management was improving my conversational skills. Growing up, I found it easy to have in-depth conversations about subjects I knew a lot about but hard to have chatty interactions filled mostly with small talk. As a result, I found talking to new people challenging. As you can imagine, getting to know people for the first time was tough.

In your initial meeting with your staff, there's a technique to make breaking the ice easy: just ask open-ended questions that allow them to do most of the talking, and let the information that you want to discover to come to the surface naturally. Think of it this way: try to keep the thought bubble over their head at all times. We'll learn more about this coaching technique in *How to Win Friends and Influence People.*

For example, try these questions:

- How's your week going?
- What are you working on at the moment?
- What was the last thing that the team shipped that you were really proud of?
- What would you say the team is primarily responsible for?
- How are your teammates doing?

Breaking the Ice

- What's your favorite thing about working here?
- What about your least favorite?
- What's the one thing that nobody wants to speak up about?
- What's the one thing that you think I could help you with?

You can use any or all of these, but they're useful if the conversation dries up or if the other party isn't naturally chatty, like I used to be.

You'll find that as you begin to spend time with your staff, you'll uncover numerous issues and conflicts within their current situation, ranging from communication to stress to interpersonal issues and beyond. These are great starting points for you to start contributing to making their experience at work better. Just make sure you're capturing the key themes using a notebook, laptop, or whatever you choose. We'll look at this in more detail in *Manage Yourself First* by introducing a system for capturing and managing your information.

End your initial meeting by thanking them for their time and letting them know that you've noted everything down. Repeat any actions you're going to take back to them for confirmation.

As you meet each of them, you'll be beginning to form a picture of how things are. Do the people in your team seem happy? Is there a mismatch between what you initially thought your team was responsible for and what they're actually doing day-to-day? Are there any patterns emerging where multiple people highlighted the same frustrations or issues?

Don't worry about doing anything right away. Just keep it written down, and let the week continue to unfold. I'm sure you've got plenty of induction activities to keep you busy.

Book a Weekly One-to-One Meeting with Everyone in Your Team

Now that you've met everyone, it's time to start putting meetings in the calendar. Depending on your background before management, you may have avoided meetings like the plague: they were the periods of the day that ruined your flow while you were focusing on solving difficult programming problems. However, you'll soon see that meetings are how a lot of your work as a manager gets done. If you have a severe allergy to meetings, then I hate to say it, but get out now. Just run away. You're going to spend a lot of time in them. But if you persist, you may learn to tolerate them, or even to love them. Trust me.

Maker's Schedule, Manager's Schedule

Y-Combinator co-founder Paul Graham wrote *an article [Gra09]* outlining the key differences between the daily schedules of makers (for example, programmers) and managers.

Managers typically schedule their days around hourly blocks in their calendar, whereas makers are most productive when they can schedule their time in blocks of half a day. This is because the work of makers is deep and takes time to achieve productivity.

As a new manager, you may be tempted to drop meetings into the calendars of your staff at your convenience rather than theirs. Sometimes it's unavoidable, but if you can, try your best to protect the time of those on a maker's schedule. Have meetings with them at times that allow them to be focused and productive during the rest of the day.

One-to-one meetings are one of the most important regular activities that you'll perform as a manager. They're so important that we've dedicated an entire chapter to them (*One-to-Ones*). So, we'll dig into the details soon, but for now, get the meetings booked in. Open your calendar software and get going.

These are some rules of thumb to abide by:

- Make them a recurring weekly meeting, ideally in the same time and place each week. Regular cadence is important.
- Make them an hour long to begin with.
- Ensure they are in a private space such as a meeting room, so both of you can talk about confidential and sensitive issues without concern that others hear.

One more thing: at this moment in time your calendar is probably looking pretty empty. After all, it's your first week. Use this to your advantage: ask each of your staff if there is an ideal time of day for them to have the meeting with you. This gives the best opportunity to not break their flow and is a considerate thing to do.

Flow

Coined in 1975 by Mihály Csíkszentmihályi, flow is a mental state in which a person is fully and enjoyably immersed in the activity that they are performing.

Also known as being in the zone, you may have experienced this highly productive state when programming, drawing, writing, knitting, fixing your bicycle, or numerous other activities that require skill and concentration.

Programmers often rely on the flow state to be at their most productive. This allows them to fully immerse themselves within the task that they're doing. This is especially important when they are debugging an intricate problem.

OK, so you've booked in a one-to-one with everyone? Excellent. We're not done yet. You need to do one more introductory meeting.

Have an Introductory Meeting with Your Manager

You've met your team individually and now it's time to do the same with your manager. But don't wait for them to schedule something in with you: be proactive and request some time with them. Don't worry about them thinking that this is rude and that you should wait for them to do so. As a manager myself, I actively enjoy when my staff proactively seek my time and guidance; it's one less thing for me to schedule in myself.

Follow the same procedure as when you did this meeting with your direct reports. Use open-ended leading questions, listen, and take notes. You'll want to be more guided in your questioning, however, since your manager will likely have some key insights that you should know about.

For example, you'll want to find out:

- Their opinion of the team and their performance.
- Details of individuals: are they superstars or struggling?
- Framing of the current project or workstream. How long have they been doing it? Has it changed much?
- Thoughts on how best to work with them. You don't necessarily have to follow this, but another opinion always helps.
- General advice about how to do your role within the organization.

Your manager's insights form another part of the snapshot that you're putting together. After the meeting, follow the same procedure as you did with your staff: thank them for their time and schedule a recurring weekly one-to-one.

So, that's the investigations over with. Let's make some sense of them.

Create Your Snapshot

Now that you're coming to the end of your first week, you'll have had your initial meeting with the members of your team and also your manager. You will have also spent time immersing yourself in the team's work, doing any required onboarding, and making some initial connections with others in the company. You'll have begun to have formed your own observations about the team: how do they work together? How happy do they seem? Are you excited to lead them or hesitant about it?

You form your snapshot by taking what you've learned from the team, your manager, and yourself and overlaying them like a Venn diagram. Then you can begin to work out where particular observations fit into the intersecting circles.

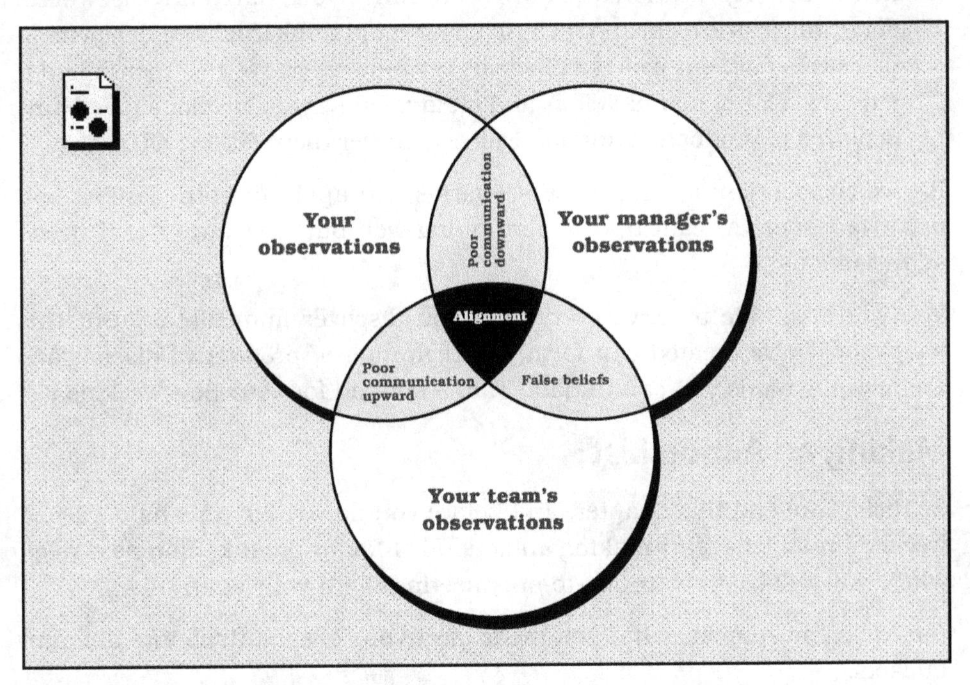

Let's look at the four intersecting sections and see what it means for an observation to fall there:

- *Alignment*: Observations or beliefs that you, your team, and your manager all share. For example, you could all observe that everyone is generally happy or that dev servers seem to break all of the time.
- *Poor communication downward*: This is where you and your manager share observations or beliefs, but the team is either unaware or disagrees. For example, both you and your manager may both observe that the team cause a high number of bugs in production when compared to other teams, but the team themselves are unaware that this is an issue. For some reason, the team hasn't been clearly informed.
- *Poor communication upward*: This is where you and your team share an observation or belief, but your manager is unaware. For example, both the team and you have observed that they are understaffed for the roadmap that they have committed to, but your manager isn't aware that they are struggling. For some reason, this problem hasn't been raised.
- *False beliefs*: Here, both your manager and the team have a conflicting observation or belief from your own. This can be especially true if you're new to the organization and have seen things done differently elsewhere. For example, the team and their manager could think that their deployment process is fine, but you can observe they only release once a month and it requires a lot of manual effort and often fails. At your previous job, teams deployed to production automatically whenever they wanted, often daily.

It's worth mentioning that these scenarios can apply to both positive and negative issues. A team that is performing well but isn't aware of it needs addressing too.

You won't be able to solve everything you discover immediately. But this snapshot you've created can form the beginning of productive discussions and investigations you can initiate. That's not bad for your first week, is it?

Making an Action List

So, before we end this chapter, and before you depart for your hard-earned weekend rest, let's start making an action list for you to take into next week where you will begin your one-to-one meetings with your staff.

Based on the snapshot that you made, sort your observations into different buckets:

- *Items to talk about with your team*: Any issues as a result of poor communication downward are good topics of conversation. Going back to our earlier example, why are there a high number of production bugs? Is it due to the QA process, or deadline pressure, or something else? Additionally, you can challenge false beliefs with your own external observations.

- *Items to talk about with your manager*: Issues that are a result of poor communication upward should be discussed with your manager. For example, why is it that you and your team can see that they are clearly understaffed, but your manager can't? Is the manager unaware that they are struggling? If so, how is their performance being measured and why is this slipping under the radar? You also can discuss false beliefs to get another opinion on the matter.

Depending on your team, the preceding buckets may have just a few items or, in the worst case, you may have generated months of discussion material. If the latter is true, then prioritize the items in order of impact so that they don't become overwhelming. Then pack your things and head on home, proud in the knowledge that you've had a very productive first week.

And Relax...

Whew! How are you feeling after all of that? Although that was a lot of work, you've come a long way in just one week. Here's what you've done:

- *Met your team*: You've spent quality one-to-one time with each member of staff and heard what their job is like from the front line.
- *Met your manager*: You've made an initial connection with your manager and heard about your team from the outside.
- *Made your own observations*: You've spent a week seeing what the job is like for yourself and others.
- *Created a snapshot*: You've made notes on what you've learned and combined them to see where there is alignment and where there are issues.
- *Booked one-to-one meetings*: You've gotten some essential admin out of the way.
- *Created an action list*: You've discovered some interesting issues to focus on next.

Without structure, many managers feel lost in their first week. However, you've excelled. Well done. Take a minute to reflect on what life was like before management. Do you feel like there's too much going on? Are you already beginning to struggle to keep your information organized? Well, keep reading: we've got just the chapter for you.

Life truly begins after you have put your house in order.

> *Marie Kondo*

CHAPTER 2

Manage Yourself First

Today started so well. You felt like you ended your first week on a high. You'd met your team and your new manager, and you'd made concise notes on everything that you were going to follow up on this week. You saw good in the world and the world saw good in you. And then it happened: a production issue that started in your team brought the whole site down. You missed several induction meetings while dealing with the fallout, then you missed your lunch. You did your afternoon catch-ups while stressed and hungry, and to top it all off, you watched in slow motion as you accidentally knocked over your bottle of water, soaking your notebook.

The doubt hits. Is this chaos what management is like? Have you made the right decision in choosing this role? Just how are you meant to feel organized and get anything done amid all of this?

Don't worry. We've all had days like this, and we'll all have them again. It happens. Catastrophe doesn't necessarily have to strike to feel like your work day was a waste of time, either. You can suffer from death by a thousand annoyances.

Perhaps you've had days where you kept getting interrupted, or you couldn't focus, or there were simply too many things going on at once to make significant progress on any of them. If you are already a manager, then I'm sure you've had one (or many) of those days. You started with a beautifully prioritized to-do list and an optimistic outlook in the morning, only for all of that to be steamrolled—urgent interruptions, issues in production, being roped into lunch with visitors—you name it, they totally killed your expectations for what you wanted to get done. You left the office wondering why you even bothered coming in.

In the previous chapter we introduced the concept of flow, which is where programmers try to spend most of their time in order to be optimally productive. A lot has been written about how to enable and maintain this state in developers, both from the internal mindset of the individual to the way in which teams build software. Take Scrum, for example—an agile framework that is commonly used in software development. Teams work on tasks in fixed timeboxes called sprints, typically one to two weeks long, contributing toward a goal. During each sprint, the team doesn't make changes that may endanger the sprint goal. New work is queued on a backlog for the next sprint. This protects the focus of the team and allows for more opportunities for our makers to achieve flow and be productive.

That's all well and good for the makers, but you're a manager now. Which frameworks and ways of working are acting in the best interests of *your* time? Is the goal of being a manager to deal with the chaos so that others can achieve productivity at your detriment?

I'm here to tell you that it doesn't need to be that way. You, too, can be a happy, healthy, and productive manager who is in control of their time. We all have bad days where nothing goes right. However, you need a methodology to rely on. In this chapter we're going to give you a system that will enable you to capture, organize, prioritize, execute, and measure your work.

This will enable you to do the following:

- *Organize yourself* so that you don't forget anything. You're never going to survive by keeping everything in your head and on scraps of paper and post-it notes. We're going to look at the system that I use to bring order to the chaos of emails, DMs, and face-to-face interactions.
- *Categorize your activities* throughout the day. Even that short conversation in the kitchen while you make a coffee, when framed in the right way, can be an impactful managerial activity. We'll introduce the categories and then show how even the most scattershot days can be meaningful.
- *Measure your output* in a way that helps you make priority decisions about how best to spend your time. We'll learn what a manager's output actually is.

Let's Get Organized!

The first part of your system that you are going to learn is how to capture and record information. As a manager, you'll find that information bombards you from everywhere: emails, DMs, face-to-face conversations, meetings, you name it. A big part of being an effective manager is to be able to hold all of this information in a place that you can easily find and review it when you need to.

Whole books have been written on effective ways of organizing yourself. However, I'm going to share with you the system that I've developed for myself over the years. You may find that it works perfectly well for you or, in time, you may find your own system. Either is fine, as long as you *have* a system. You won't survive without one.

The core principle of the system is this: *you want to keep as little information in your head as possible.* Human brains are terribly clever, but when there are many sensory inputs, they're not great at remembering every little task. How many times have you said to yourself that you'll pick up dishwashing liquid on the way home from work, only to reach for the nonexistent bottle after dinner? It happens because we are humans. But we've invented many other ways for capturing and organizing information, so let's use them to our advantage.

This system boils down to four tools that form your tool belt:

- Your calendar: for organizing your *time*.
- Your to-do list: for organizing your *tasks*.
- Your email inbox: for organizing your *incoming messages*.
- Your place to capture information: for when you're not in front of the other three tools.

All of these tools are essentially free and easy to use. The bonus is that if you are able to master this system, not only will your work day feel calmer and more under control, other people will think that you're a *wizard*. You'll be able to remember that one thing you said you'd do for them when you chatted to them in the kitchen last week. I've never written a line of code that has impressed people more than my ability to remember things because of this system. Getting the hang of this will make you seem like a managerial superhero—The Great Managerius, if you will.

Let's now have a look at each of the tools and how to use them. Afterward, we'll look at what an example day looks like when using this system so you can get a feel for what it means to put it into practice.

Tool 1: Your Calendar

Love them or hate them, calendars are going to be a big part of your life as a manager. Your calendar is the de facto record of what you are doing with your time. You should live your day by your calendar. Every meeting that you need to go to needs to be in there, otherwise how else are you going to remember to go? You should also ensure there are plenty of free spaces in your calendar that others are welcome to book if they wish.

So far, so good. Since there's a strong possibility that you already work in an organization that uses calendar software, it's likely that nothing in this paragraph has been much of a surprise. However, you can use some additional techniques to keep your calendar as organized as possible.

Firstly, only use your calendar for organizing your time. Your calendar is not your to-do list. Keep tasks off of your calendar and use a to-do list instead; to-do list software is much better suited to that kind of thing, and it prevents your calendar from looking like a panic-inducing jumble.

Only put the following things in your calendar:

- Meetings with other people.
- Activities that indicate that you are busy (for example, lunch, out of office, doctor's appointment).
- Blocks of time to prevent others from booking over them (for instance, when you need to focus).

And that's it. If those are the only things in your calendar then it might get busy, but it will be sequential and organized. At a glance, your calendar should show where you are currently and where you're going to be later. Make your calendar easy to interpret by not overloading it with anything else. It's also considerate to others: people can look at your calendar and see when you're free. They don't have to look through each individual entry and decipher whether it's an actual meeting or a reminder for you to send out an email. Keep your tasks elsewhere. Tidy calendar, tidy mind.

Your calendar is as much for you as it is other people. However, being a manager is not to have your time controlled entirely by others. Allow yourself periods of the day where your time is not to be interrupted by meetings. Block out regular chunks of time to focus. Simply put an entry in your calendar, usually of a few hours in length, called "no meetings please." Those are the times when you can close your email, put on your headphones, and concentrate on a task. It's well within your right to do so.

Next, remember that even though your calendar may have nine hourly slots for a working day, your energy is not going to be consistent across all of those slots. Ensure that when you are managing your calendar that you allow yourself time to grab lunch every day and to get away from your desk and into the fresh air. For example, during a period of heavy recruitment, a lack of calendar management could mean that you get booked into eight hours of back-to-back interviews with no break, and that will be nobody's fault but your own! I recommend taking the approach of having your meetings being public (to the organization) by default and only changing the privacy if the content of the meeting is sensitive. This helps reduce the mystery to others of what you are spending your time doing. Transparency is a good thing.

Lastly, make sure that you use reminders. Take advantage of automatic notifications that remind you ten minutes before the next meeting starts. Set them up on your computer and your phone. This makes sure that you don't forget about where you're going if you're away from your desk. They also serve as a good reminder to begin wrapping up the current activity before moving on to the next one.

Open your calendar as your first task at work every morning and review the day; this allows you to prioritize any preparations if necessary. Get in the habit of shifting meetings around to batch them together to give you a chance at achieving flow. Before getting on with your day, protect periods of free time with "no meetings please" blocks that allow you to work uninterrupted.

Rules: Calendars

Calendars are extremely useful, but they can become perilsome if you use them incorrectly. Remember the following rules:

- Your calendar is only for organizing your *time*. Nothing more.
- It is your responsibility to manage your calendar, including ensuring that you get enough breaks and deep work.
- It is both for you and other people to use. Keep it tidy and meaningful. It represents you.
- Making meetings public by default can help others reason better about how to schedule time with you.
- Use notifications before the next meeting to make sure that you don't miss the next activity.

That's your first tool of the trade: your calendar. Be disciplined, manage it, and manage yourself through it. Create structure for your day. Actively prune it and shift it around where needed to have it support you, rather than tell you what to do.

Your Turn: Your Calendar

Before we move on to the next section, have a look at your calendar and ask yourself the following questions:

- Do you currently use your calendar for anything other than just scheduling your time? If so, what? Where should that be recorded and actioned instead?
- Do you actively sit down daily or weekly to review the coming week and move your calendar around to better support you?
- Do you block out deep work time in your calendar? Would you be able to do so regularly?
- Is there anyone that you work with that has a crazy calendar? Have a look at it. Why do you think that is? (The worst calendar I have seen was a combination of work calendar, to-do list, home calendar, and reminder system. It filled me with existential dread. I have never forgotten it to this day.)

Tool 2: Your To-Do List

So we've looked at calendars, which are how you manage your time. But how do you organize and prioritize your tasks? That's where your *to-do list* comes in. I'm sure you've used to-do lists before. But, like calendars, a certain amount of discipline is required to make them really work for you.

Your to-do list should be the *only* place where your tasks live. Tasks may temporarily live elsewhere, such as your brain or in an email, but ultimately the to-do list is their home.

Let's suggest a strict rule for yourself: if it's not on your to-do list, then you're not doing it. This has two benefits: it makes you keep your to-do list up-to-date and prioritized, helping you stay organized, and it also allows you to free up mental space through not worrying about all of those other tasks that you need to remember to do. What should you do with an email asking you to do something? You put it on your to-do list. That Post-it note someone left on your keyboard? Guess what?—it goes on your to-do list.

Your to-do list is your launchpad for each day. It contains all of the things, no matter how mundane, that you should do. You can drive yourself from it. It should be the first thing that you open in the morning and the last thing that you close at the end of the day.

For those of you that aren't used to running your day from a to-do list, then it does take some practice and discipline. However, you'll learn to love the structure that it brings.

To-do list software has some killer features that make it hard to ever consider going back to paper. For example, many to-do list applications allow you to have recurring tasks that can pop up on your list at a regular schedule such as daily, weekly, and biweekly. This allows you to automate reminders to yourself to help keep as few things in your brain as possible, which frees up time for you to think of other important things.

For example, I have daily recurring to-do items to check my email and view and organize my calendar as the first thing when I get to work. This might sound silly, but it helps me form habits. I also have weekly recurring tasks that remind me to prepare for my one-to-one meetings on the day so I can come prepped with an agenda. I have similar weekly recurring items for various weekly digest emails I send to the company and also my own manager. By allowing my to-do list to automate all of my reminders, I can just calmly follow along with what my trusted partner in crime says I should do.

So your second tool of the trade is the to-do list. Used with discipline, it can make others think that you have superhuman memory and organizational skills, but in reality it's just some simple software used with simple rules.

Rules: To-Do Lists

Here's your rules for to-do lists:

- Use whichever software works for you. However, until paper can sync with multiple devices and automatically repeat recurring tasks, software to-do lists are your best bet.
- Make your to-do list the first thing you look at in the morning and the last thing you look at before you go home at night. You'll soon learn to love the art of checking off items.
- Use it to remind yourself of anything you want: to check emails in the morning, to tidy your assigned tickets, to send that thank-you email. Keep as much out of your brain as possible.
- If it isn't on your to-do list, don't do it until it is. It'll force you to form a habit of using it.

Which Software Should I Use?

You're beginning to feel more organized. You might even be looking to try out different to-do list software. That's great! If you're looking for somewhere to get started, then the software that I use myself is Asana. There may be others out there that are better suited to you, but I've come to rely on the following features:

- *Recurring to-do items*: So that you can automatically remind yourself to do things.
- *Due dates*: Each item can have a due date so you can prioritize and schedule things how you wish.
- *The ability to hide information*: The software lets you put anything not due to be completed today into Later, which can be hidden from view to reduce the feeling of information overload. As time passes, tasks automatically reappear in Today.
- *Subtasks*: Each to-do item has a free-text description box, which is handy for writing long form notes in, and also allows subtasks and comments.
- *Categories*: You can put to-do items into categories, which allows you to show and hide different projects and streams of work from view. This is good for controlling information overload.

Using these features has made me happier and more organized. There's even a feature that sends a little animated unicorn flying across the screen when I complete a task. Excellent.

Which to-do list software do you use? Give some of them a go. Similar to picking IDEs for programming, the choice is highly personal. If you're feeling brave, you could even use emacs org mode!

Your Turn: Your To-Do List

Consider these questions before moving on:

- Which to-do list software looks like the best one for you?
- Do you currently keep to-do tasks in multiple places? If so, where?
- Do you also use emails as implicit to-dos by leaving them unread or in your inbox? How often do you forget to action them?
- Try out a day that is driven from your to-do list. How did it feel? Did you begin to notice that you could operate with a trust that you hadn't forgotten anything, allowing you to focus more on what you were doing?

Tool 3: Your Email Inbox

Let's look at your third tool. It's email and it makes the world go round. To struggle with email—and to go so far as to hate it—means that you're resisting the irresistible. Instead, you need to be able to master a reliable system for making your email inbox work for you, and not the other way around.

Learning to love email (OK, you won't *love* it, but it's as much part of your life as brushing your teeth) involves implementing some simple processes, but mostly it involves a mindful approach with some self-discipline.

In the same way that you've probably seen a colleague with a crazy calendar, you've probably also glanced at colleagues' screens and seen their crazy email inbox. Tens of thousands of unread messages! Emails to themselves as reminders to do things! Hundreds of emails from newsletters that they *might* read! My blood pressure is rising just thinking about it.

Rules: Email

Here are your rules for email:

- *Your email inbox is your primary source of incoming messages.* Any other software that you use, such as your chat application or ticket tracker, consider to be a secondary source. Configure all other pieces of software to email you if someone messages you or if there's activity in other applications. This means you only need to be on top of your email, and everything else will follow.

- *Archive anything you're done with.* If you've read it, archive it. If you've replied, archive it. If you've got to do something other than reply, put that task on your to-do list and archive the email. This way your email inbox only ever has a few emails in it at any given time. If you need to find something again, you can just search for it. It might not be in view, but it's there.

- *Never delete email.* It is a transactional record of communication. Everything should be archived, but never deleted. This means you can search for that conversation you had in 2010 and still find it. There will always be that email you need to find five years later.

- *Unsubscribe from lists you don't read.* If you sign up to that newsletter but end up not reading it, just unsubscribe. Never stay on mailing lists you're not engaged with. It just creates clutter.

- *Set up an email rule for important messages.* If you get a lot of email, then you can tell your staff that including a particular word in the subject of the email will automatically filter it to the top of your list. This can allow them to be seen quickly amid the chaos.

Although these are a few simple housekeeping items, they're extremely powerful at keeping your inbox under control, especially as a manager where you'll probably be getting a whole load more emails than you used to.

You should perform multiple batch email sessions a day rather than keeping your email open all of the time. Email is asynchronous by nature, so take advantage of that. Don't be an email inbox watcher, as you can spin in a busy-wait pattern and get nothing done: protect your focus for getting real work done.

Inbox zero is an extension to the approach above, where you aim to keep your inbox empty at all times. Depending on how much email you get, this may or may not be a good use of your time. Certainly if I had to choose between stopping work at the end of the day and answering a bunch of low-priority emails, I'd rather stop work and come back to them tomorrow. Don't beat yourself up over inbox zero. Use it as a guide rather than a doctrine.

One important item is worth reiterating: email is *not your to-do list*! Your to-do list is your to-do list. Put actions that you need to perform on there and archive the email relating to it. I cannot stress that enough. From experience, this is the primary reason that people forget to do something when you email them. They'll read the email, make a mental note to come back to it later because they're doing something else, then forget about it among all of their other emails. It's extremely easy for this to happen. Take an action, put it on your to-do list, then archive the email. Job done. Don't be that person.

Your Turn: Your Email

We've got one more tool to go. But before we move on, here are some questions for you:

- Think about how you currently manage your email inbox. Do you manage it at all? Do you have a system? Is it working for you?
- Spend a day following the rules set out above. How did you feel doing this? Was it uncomfortable archiving if you don't already do it? What did you think about having a small inbox?
- Do you overload your inbox with other tasks that aren't just email? Are you also using it as a to-do list, reminder system, or something else?
- How many unread or ignored mailing lists are you subscribed to? How many could you unsubscribe from right now?

Tool 4: Your Place to Capture Information

Just one more tool remains, and this is the one that you have the most flexibility about how to implement. If you are sitting at your desk, then you only need the first three tools that we've outlined to do a great job. However, there's a missing piece of the puzzle: what do you do when you're away from your computer and you need to capture a task for yourself? For example, that quick conversation in the kitchen may result in you needing to remember to do something. Since we don't want to have to commit anything to memory where possible, what's the best portable way in which you record information?

Personally, I tend to use my phone and I jot things right into my to-do list. But you may find that a small notebook works best for you. Whatever it is, ensure that you've got something on you at all times that you can capture information with while on the move, or whenever an interesting thought occurs. Then all you need to do is to transfer anything you've captured into your to-do list the next time that you're there.

- What's the best place for you to capture information? Is it a notebook, phone, or something else?

- Using your phone to jot notes in a meeting might be seen as rude as others may think you're ignoring them. Be mindful of this.

Your Toolkit: A Summary

That's it: you have the only four tools you need to stay sane and never forget anything important. Before moving on, let's review what we've learned. These are the four tools:

- Your calendar: for organizing your *time*.
- Your to-do list: for organizing your *tasks*.
- Your email inbox: for organizing your *incoming messages*.
- Your place to capture information: for when you're not in front of the other three tools.

Trust your tools. Use them daily. Follow the simple rules. You'll be surprised at how much you can remember, how productive you'll feel, and also how your colleagues will think that you are one of the most organized people that they've ever met.

An Example Day Using Your Toolkit

Let's have a look at an example day using your tools.

8:45: You sit down at your desk and open your to-do list. Today already contains some recurring tasks, such as preparation for a weekly meeting and a one-to-one. It also contains a daily reminder to go through your emails. You open your calendar and see what the day holds. You add a few items to your to-do list. You then prioritize the list by dragging the tasks into a different order. You tell yourself that if you can get through that list today, you'll feel accomplished.

9:00: You start going through your emails. Some require no response and get archived; they are purely bulletins. Others that do are replied to and then archived. Anything requiring action beyond a response—for example, having a conversation with someone—goes into your to-do list. Then you archive the email.

9:15–12:00: You go about your morning. You have a couple of meetings where you take short notes in your notebook. You also note down some reminders to yourself in the same place. When you get back to your desk, you enter these as to-do list tasks and reprioritize them and begin working through the actions.

12:45: You go through emails once more, archiving them as you go. You now have zero emails in your inbox.

12:50: Ding! Your calendar reminds you to go and grab lunch.

1:50: You're back at your desk for an hour before a weekly steering meeting starts that requires some preparation. This is the most important current task on your to-do list. You do that preparation and head off to the meeting.

3:00: You bump into a colleague in the kitchen and she recommends you check out a new open source project that's gaining traction. You jot it down in your notebook.

4:00: You're back at your desk and you translate your recent notebook scribblings into your to-do list. These get labeled with tomorrow's date, as they're not urgent. You put them into Later so that they disappear out of your sight until tomorrow morning. You then get your head down and get as many remaining tasks done as possible. You're determined to finish your list by the end of the day.

5:30: They're all done apart from one, which was the lowest priority anyway. Oh well, you'll kick that to tomorrow by setting the date on it and moving it into the Later section. Your list for today is empty. You close the tab.

5:45: One more pass on emails, archiving as you go, and you have an empty inbox. Home time!

Simple habits repeated frequently lead to a clear mind. Use this system for getting things done and you'll begin to wonder how you managed without it.

How to Categorize Your Activities to Feel Productive

Regardless of how organized you are, various forces act against your personal feeling of happy productivity. As a programmer I could look back at the code that I'd written at the end of the day, point at it, and say "I did that." I could also point at our ticket tracker and say "I did those tickets." Even on frustrating days where there was a production issue or something gnarly to debug and fix, we would eventually solve the problem, commit the code, and again I could point at it and say "I did that!"

Several reasons explain why it's often hard to find this satisfaction as a manager:

- *You're primarily working through other people*, so there's less to feel that you have tangibly done by yourself.
- *You do dozens of little tasks* rather than fewer substantial ones.
- *You are context-switching throughout the day*, which can be tiring and frustrating.
- *More of your time is spent in meetings and discussions*, which do not produce a concrete document, block of code, or production deploy.
- *You're more likely to be interrupted* and just end up not being able to get your work done anyway!

If you're unable to frame all of your managerial work in a way that is able to make you feel productive, then, with time, you may feel like your job is happening around you rather than feeling like you are doing your job.

There is, however, an answer. One of my managerial heroes is the late Andy Grove, who was one of the founders and long-term CEO of Intel. His *advice [Gro95]* for tackling this problem was to categorize the managerial activities that you partake in daily into four buckets:

- Information gathering
- Decision-making
- Nudging
- Being a role model

If you apply these categorizations to your activities within the day, then you'll see how there's more meaning in your work than you may originally think. You can also keep these categories in the front of your mind to better choose how to spend your time.

Let's have a look at them in turn.

Information Gathering

As you may have gathered from the need to have a system to capture, record, and action tasks, which we introduced at the beginning of this chapter, the information that you hold is critical to your success as a manager. Your knowledge base is what you use to understand what is going on in your team and the wider company and is fundamentally what you base your decisions on. The activity of information gathering feeds this knowledge base.

It's worth noting that information gathering isn't a formal process. It can happen anywhere at any time. For example, you could be having a conversation at the coffee machine with your colleague when she mentions that her team is building a new API as part of another of the company's products. You realize that this could be helpful for your own team, so you note it down. Weeks later, your product manager asks whether your team could build this feature yourselves. You already know that an API exists, so you make the connection between the teams, and you save your own team a lot of work.

Keep adding information to your knowledge base. Seek it out, ask questions, and be inquisitive. The desire to gather information will also motivate you to talk to others, make connections with your peers, and be a better manager.

Decision-Making

This is one of the more obvious answers to the question "What does a manager do?" You can make decisions of all sizes. These range from small, such as granting a holiday request, to large, such as deciding whether to migrate your application infrastructure into the cloud or keep it within your own data center.

Do take decision-making seriously. The ability to decide is a privilege that not everybody has. It's easy to forget that there are many people who don't have the power to decide particular outcomes, so always give decisions your full attention and take responsibility for the ramifications of making them. Every decision is an inflection point: should we hire Bob or Susan? Should we split the team into two sub-teams? Should we refuse to begin estimating the work required for this project when the proposal for the product is so unclear? Decisions such as these may seem like they're small in that moment, but extrapolated over time and bringing in the cost of the different outcomes, they are big decisions, so treat them with respect.

Nudging

The concept of nudging is influencing a decision by contributing your own viewpoint to the discussion. For example, you may be involved in a discussion about whether to build or buy some particular software, and you make it clear how you feel about the situation. You are not the decision-maker, but you can influence the decision. Like decision-making, nudging can occur for decisions of all sizes. You may put your viewpoint across about whether to book a meeting immediately or tomorrow, or equally state your case in a discussion as to whether to open an office in the United Kingdom or abroad.

Try to view your daily interactions through the lens of nudging, and you'll soon see that there are ample opportunities to broaden your influence on the organization, thus increasing your output as a manager. Like information gathering, this can also happen anywhere and any time. By having a managerial position, you also need to be aware that your word carries some authority, so bear that in mind when offering your opinion freely. This is especially true the more senior you get. Have you ever seen an offhand remark from the CEO turn into a real project even though she may not have wanted that to happen?

Being a Role Model

Being a good manager is about walking the walk as well as talking the talk. The best way to demonstrate to your staff and your peers is to lead by example. Be present and visible, get involved in day-to-day discussions, and contribute technically if you have the time and inclination. Demonstrating the standards that you wish to see others perform to is the best way to create change, so lead from the front! If you want to increase the flexibility your team has in their core office hours but you're in the office from 8:00 a.m. to 7:00 p.m. each day, then you're not being a role model. Go home!

An Example Day Categorizing Your Activities

Let's have a look at a reasonably typical day and see how we can categorize seemingly small and unimportant activities into opportunities to do managerial work.

8:45: You sit down and prioritize your to-do list. You read your emails and unread chat messages. Here you are *information gathering*. What's going on? What do you need to do? How can you contribute?

9:00: You answer your emails. You contribute to various discussions with your viewpoint, which is *nudging*. You decide to make an offer to a candidate you interviewed yesterday. That's *decision-making*.

9:10: While in the kitchen and making a tea, you have a conversation with a colleague and learn what they're working on. *Information gathering*. You share how your own team tackled a similar technical issue. You suggest taking a similar approach. *Nudging*.

10:00: You attend a meeting to review a number of resumes that have come in over the last few days. You choose which to invite to a first interview. *Decision-making*. You suggest to the CTO that it's a good idea to open the position out to more junior candidates now that the local universities are a few months away from having large numbers of students graduate. *Nudging*.

11:00: You're in a one-to-one with a direct report. You are *nudging* them, as ideally you want to steer them into making their own autonomous decisions. You learn a lot about what the report has been working on this last week and how the issues have been overcome. *Information gathering*. You offer some opinions of how you might tackle the next piece of work. *Nudging*.

12:00: Lunch. You gather some food, rather than information, at this point. But while eating you do have a conversation with a colleague about his experiences using Jenkinsfiles, since your team has moved across to using these recently. You give some advice about who to talk to for them to learn more. *Nudging*.

12:30: You catch a colleague in the breakout area who shipped some new functionality last week. You tell them that their team did a brilliant job and that the user feedback has been great. You do this because you want your colleagues to get better at delivering honest feedback. *Being a role model*.

1:00: You go through your emails and messages, both *information gathering* and *nudging*. You have a decision to make about whether some work should be put into your team's backlog or not. You decide that you need to talk more in person about the feature, so you set up a meeting for later.

3:00: You have the meeting. Your product owner describes how the work can make your own product more compelling, and you also know that you have the technical expertise to build it in such a way that other teams can use it too. You both decide to take the work on because contributing to other teams as well as your own is a good example to set. *Decision-making* and *being a role model*.

4:30: You spend the last couple of hours in the quiet going through items on your to-do list. One of these items is preparing an engineering talk on your latest project (*being a role model*). At the end of the day, you read your email (*information gathering*), review some pull requests (*decision-making*), and take part in a discussion in the back-end development channel about best practice around log aggregation (*nudging*).

When categorizing your day it's possible to see how even fairly mundane interactions can be transformed into an opportunity to be a force for good and improve the organization that you work in. Also, when a day goes totally down the pan, you can *always* retroactively categorize elements of it as worthwhile, even if you didn't make much progress on your to-do list.

Your Turn: Your Activities

So what are you waiting for? Try it out. It's fun:

- Spend a few days noting down your activities at work and then categorizing them into the four activities. How much time do you spend on each? Do you neglect any of them?
- Go into the proceeding days more consciously thinking about your work and these categories. Do they make you act differently in situations?

How to Measure Your Output as a Manager

So far, so good. We've looked at how to organize your daily activities and also how to categorize them into activities so that you can better reason about what you spend your time doing and feel more productive. Being able to articulate those activities allows you to measure them. But what actually is a manager's measurable output?

The best rule of thumb that I ever learned can also be attributed to Andy Grove. He states that a manager's output can be measured by the following *equation [Gro95]*:

The output of your team + The output of others that you influence

This is simple but powerful: it can be used to make decisions about how to spend your time. We'll be using this equation as a tool as we learn techniques for working with individuals in the next chapter. But think about it for a minute. You may begin to see what framing your output in this way means:

- You are ultimately accountable for your team's output. If they're not able to perform well, then it's your job to improve it.
- Your team's output is more important than your own personal output. This means you should be spending more time delegating, coaching, and mentoring, rather than doing tasks yourself.
- You also do your work through those outside your team that you influence. This means all of that nudging and being a role model is important. It helps improve the rest of the organization too.

It comes as no surprise that a scientific measurement of a person's output is nigh on impossible (have you ever been subject to debates about measuring the productivity of programmers?). However, having this equation as a mental model to use when deciding how best I should allocate my time has time and time again made me a better manager. It's now yours to use too.

Your Turn: Your Output

Look at the equation again, and then do the following:

- Think about your previous week at work. With the equation in mind, would you have made different decisions about how to spend your time? If so, why is that? If not, why did your activities align with the equation?
- Visualize the best manager that you've ever had in your career. Can you reason as to why they were the best by using this equation?
- Keeping that manager in your mind, are you able to reason about their output using the four categories that we defined previously?
- Now think about the worst manager you've had. Oh go on, you don't need to tell anyone who it is. Why were they a bad manager in relation to the equation and categories of activities? What were they not doing?

OK, I'm Ready!

So there we are: you've learned a lot in this chapter.

- Firstly, we looked at how to *organize yourself* so that you have the reliable foundations of excellent organization to build upon.
- Next, we learned how to *categorize your activities* so that you can better reason about the work that you're doing and also see how an ad-hoc chat or an interruption can be turned into a valuable managerial activity.

- Lastly, we presented a way for you to *measure your output* to ensure that you're always thinking about how best to allocate your time to be the most effective manager that you can be.

This brings us the end of the first part of the book. Well done! However, we've only just begun to scratch the surface. So far, you've learned the tools that you'll need to get oriented and manage yourself. But management fundamentally isn't about you, it's about facilitating others to succeed. We're going to take a bold step into Part II, which is where we encounter the scary stuff: working with other people. Don't worry though, we've got you.

Part II

Working with Individuals

An essential management skill is being able to work with, and through, other people.

However, if you thought that getting computers to do the right thing was hard, wait until you start working more with humans!

We'll be covering everything to do with people: how to communicate, do one-to-ones, how to motivate and praise, how to do performance reviews, hire, fire, and expand your network.

Once you've got the right APIs, it all gets a lot simpler.

The single biggest problem in communication is the illusion that it has taken place.

— George Bernard Shaw

CHAPTER 3

Interfacing with Humans

You're standing in the kitchen thinking about what to make this evening. You're flicking through a recipe book full of bright pictures and sequential methods. Recipes are sort of like algorithms, aren't they? You set up the environment, follow the steps, and out comes the result. They appeal to our minds as engineers: creativity, precision, methodology, and tastiness. As you think about algorithms, your mind drifts to when you were first learning to program. It wasn't a passion that you already held, nor was it one that unfolded in an instant. It unfurled slowly with time. At first, you hated it. It was frustrating. You didn't understand any of the concepts or familiar building blocks that underpin most programming languages: variables, methods, arguments, and loops. The compiler spat out incomprehensible error after incomprehensible error as you tinkered.

But gradually, and with effort, you learned how to interact with the computer. You remember the first program that you wrote on your own. It was part of a homework exercise. It asked the user to enter a number n and then it would print out the sum of 1 to n. You felt a wave of satisfaction when it worked, instantly outputting the result to the console.

You remember in the following years how you spent countless hours tinkering with programming projects, both inside and outside of school and university. You downloaded open source software and marveled at those humongous codebases written by other people. You poked around in them to see how they worked. Your final-year project on your degree was a compiler for a small, fun programming language that you invented. And then, the impossible seemed to happen: you got a job that allowed you to get paid for your hobby.

Yet, here you are, having made a drastic decision to stop being a humble programmer that writes code. You are now a manager. You feel like you did sitting in front of that blank text editor for the first time, wondering how on earth you were going to write a program. You didn't know how to work with the computer, how to speak to the right language, how to issue instructions, and then make it execute and get the right result. What exactly is the language for running a team? How do you express what you need from them, and how do you get the output that you desire?

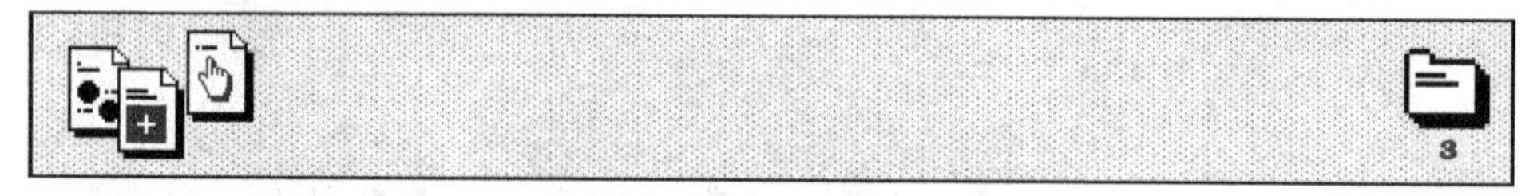

Starting a managerial role involves learning a new set of tools. You may have ended up working in technology by tinkering with the tools required to program a computer. These tools may have been programming languages, software for creating graphics, or commands for configuring servers and data centers. As your expertise in these tools increased, you became more effective, knowledgeable, and able to tackle larger and more complex problems. You were also able to teach others how to use these tools so that they were also able to become more proficient themselves. You were even able to get paid for doing so. As a manager, you may still incorporate your existing expertise into your new role, but now there's a new set of tools to learn. These are the tools that allow you to be an expert manager, which involves doing your work effectively through others.

Let's recall the equation that we outlined in *Manage Yourself First*. Your output as a manager is:

The output of your team + The output of others that you influence

This chapter will introduce the tools that you need to supercharge your output. You'll learn how to communicate candidly and concisely with your team and your manager and how to delegate work to your staff.

When first learning how to program computers, you're faced with a number of unfamiliar tools. It takes a significant investment of time to learn enough to write a program that prints out "Hello, world!" to the console. For the experienced programmer, learning a new language is easier because they already understand the underlying concepts, and they can map what they have learned in one language to another and understand the differences and similarities.

When a programmer gets their first job, they will often have had experience in using the tools that they need to be productive. For example, a junior will have almost certainly done some programming before, either as a formal part of their education or through their own self-directed learning. The same is true of a programmer moving from one company to another. They will likely be working in an entirely different domain, but they'll bring with them their expertise in their tools, which will allow them to hit the ground running and not have to learn everything all over again.

Managers in the technology industry have a tooling crisis. Typically, teams end up being run by engineers being promoted into management positions, and depending on the organization, they may get excellent mentorship and training or none at all. This results in a wild disparity in the skills of managers in our industry. Some are taught the tools that they need to become experts at their job, whereas many are metaphorically sat in front of the text editor with no idea what to type.

If we were to pick an arbitrary piece of software that we wanted to understand, we may ask:

- What is the *language* that the program is written in? How is it structured, and what syntax does the programmer use to express what they mean?
- What is the standard *API* for that language, and how should it be used to make the program do common tasks in a predictable and efficient way?
- How does the program *run* when it is written? Does it execute on a virtual machine, or is it compiled directly to run on the local operating system?

By using some artistic license, we could reason that you need some related concepts to be able to work with a team for the first time:

- What is the *language* and communication style you should use so you can discuss, debate, and deliver feedback both openly and candidly?
- What is delegation and how should you use it like an *API* for getting work done through others? How can you use delegation to get tasks done efficiently and to a high standard?
- How well does your team *run* and how is it doing in the context of the department and organization? We look at the tools for how to discover this in partnership with your own manager.

That's exactly what this chapter will be about. Firstly, we're going to learn all about communicating effectively with people. Then we're going to learn about delegation, which is how you get your work done as a manager. Then we're going to focus on the often-neglected relationship with your own manager, teaching you how you can get the best out of it to ensure that both you and your team are doing a good job in the context of the wider organization.

How to Communicate Well

Good communication is the connective tissue between everything you do as a manager. It can make all of your activities more effective. From daily interactions with your staff, to answering emails and direct messages, to performing your one-to-ones, to giving performance reviews, to interviewing candidates for a role on your team: if there's one skill that you can improve that will make all of these activities better, it's *communication*.

Most importantly, good communication forms great relationships between you and your staff. These great relationships form strong teams. Being able to communicate clearly, candidly, and empathetically will make you a manager that others want to work for. If you demonstrate strong communication skills, then others will experience it and—more importantly—replicate it. It's an excellent example of how being a role model can work in practice.

The Three Mediums of Communication

Broadly speaking, there are three ways in which you'll be communicating with people at work. The first two are performed with explicit action. The third is subtler. However, all three are worth your effort in consciously improving. They are:

- *Spoken* communication, encompassing everything from formal meetings through to conversations around the coffee machine.
- *Written* communication, which covers formal letters, emails, pull request comments, and instant messages.
- *Nonverbal* communication, which refers to gestures, body language, facial expressions, posture, and tone of voice.

Usually, spoken and written communication are grouped together with the term *verbal* communication. However, it's useful to separate them as there are some specific considerations for each of them that we'll cover.

Before we dive in, remember this key point: try your best to represent your best self no matter how and where you are communicating. In the same way programmers continually strive to write elegant and efficient code, managers should strive to be elegant and efficient communicators. This might mean you need to work on your speaking skills in order to match your writing skills. Maybe it means you need to pay attention to your body language in meetings.

Choosing the Right Medium

The world of work unfolds across many different platforms. For example, in addition to face-to-face communication in the physical world, there's also video conferencing. You will also communicate through emails, chat applications, and a myriad of other tools such as project management and ticket tracking software.

For example, on a given day, you may find yourself communicating via:

- Face-to-face communication
- Video conferencing software
- Phone calls
- Emails
- Shared documents
- Chat applications
- Project management software
- Ticket tracking software

And the list goes on. Each of these mediums have unique traits that make them good for some things and bad for others. You need to be aware of the medium that you're using, and use the most appropriate one for the message that you're trying to relay.

As a manager you'll want to:

- Choose the right medium in which to communicate your message.
- Communicate effectively using the inherent traits for that medium.

But what does that actually mean?

- *Face-to-face* communication is best for exchanges that require real human connection, where you can bring your physical presence and use your body language. However, it's synchronous and requires both parties to be there. It's best for important conversations, such as delivering a performance review. It's also the hardest to plan and steer, since you don't have time to contemplate long and hard over your replies. Sometimes this can work in your favor though, such as when a one-to-one takes an unexpected excursion into a meaningful topic.

- *Email* is fairly formal. It is also archival, which makes it the perfect place for information that you'll want to refer back to. The medium has an expectation that it's highly asynchronous: if you don't reply to somebody's email today, it's unlikely that they're going to be offended. You can use this trait to your advantage.

- *Documents*—especially collaborative ones—are excellent for formalizing ideas such as the design of a new piece of architecture. You can take full advantage of all of the features of an office suite.

- *Chat* carries an expectation of synchronous exchange; especially on platforms where it shows you who is online and typing. It's great for communication in teams, informal exchanges, and asking quick questions. It's not always a good place for formal discussions because it moves quickly and the reader can easily get lost.
- *Use-case specific software* such as ticket trackers or project-management software are excellent for solving a specific workflow problem—such as organizing your current sprint—but often have subpar functionality compared to the tools above.

When you prepare to communicate something, think carefully about the medium that you should use. It can make or break the delivery of your message. For example, you'll want to do one-to-ones and performance reviews using face-to-face communication, because the human connection and use of body language is important for forming relationships and delivering difficult messages. For a weekly update to your manager, you may choose email because it's archival. Email may also be best for confirmation of a salary increase, but the news may be best first delivered face-to-face.

When communicating, first think about the message you want to deliver. Then put yourself in the recipient's shoes and consider how best they might like to receive that message. Then adapt it to that platform. Don't take two hours to write a lengthy and complex email when you could just talk face-to-face instead.

Lastly, ensure that each piece of communication that you do is fully formed and either is understandable on its own or has a call to action. There's a big difference between this message:

```
Hello
```

And this one:

```
Hello, are you around today?
I'm looking for some help with testing my API change locally.
```

Managing Your Energy

In addition to choosing the right medium in which to speak to people, you need to be aware of your energy and how it manifests in your communication. This isn't a pseudoscience; it's real. As a manager, you'll be in all sorts of conversations. Some will leave you feeling happy, some sad, and some angry and frustrated. We're not managerial robots. We're humans. So if you've just come out of a frustrating meeting that has made you want to throw your laptop out the window, then be aware that you'll be carrying this frustration into the next meeting, email, or chat message that you write.

This might mean that you end up making somebody else feel that you're annoyed at them, when in fact you're annoyed at somebody else. Even worse, you may transfer this frustration and anger over to them! So you need to be mindful of your own mood. Check in on yourself regularly throughout the day. How are you feeling, and is this how you want to present yourself to others? Be aware that even if you are an expert at hiding your frustration in your spoken communication, the person having a conversation with you can read how you feel behind your furrowed brow, folded arms, and tense posture.

You can deal with residual bad energy in two ways. Both are equally acceptable:

- Tell the next person that you're speaking to up front about how you're feeling and, if you're able to share, why that is. They'll immediately know that it has nothing to do with them, and they may even offer support.
- Delay the next activity that you're doing until you've become more centered. If it's a meeting, just be honest. "I'm really sorry, but I've just had a really frustrating meeting and need to get fifteen minutes of fresh air. Can we start then? I want to be able to focus properly on our conversation."

Measure Twice, Cut Once

In carpentry there's a proverb: measure twice, cut once. You can't uncut some wood, so you want to make sure it's correct before doing so. Do it wrong, and you'll need to do it over again as it won't be fit for purpose.

The same is true for communication. As a manager your word carries weight: it can be interpreted as a command or as feedback. Therefore, you should think twice before you broadcast information.

Have you ever gotten some exciting information at work, felt the sudden urge to tell somebody else immediately, and did so, only later for it to be totally untrue? Have you ever had somebody tell you that they've obtained some information that you should know, but they can't tell you? Frustrating isn't it? Don't be the manager doing that to other people.

Don't communicate when you *want* to. Communicate when you *need* to. For example, if you've heard a rumor about the next project that your team is going to work on but it isn't confirmed yet, is it absolutely necessary that your team knows that rumor? Not really. Wait until it's confirmed, thus making your communication concise, clear, and trustworthy. Don't spread gossip. A manager's communication should never be driven by increasing their social capital. It should be informative, useful, and actionable.

Measure twice, cut once.

It's Not About You

Communication happens between yourself and others. However, how you communicate should never be about you. It should be about *them.* When you're writing that email, or having that conversation, try to picture what you are delivering from the viewpoint of the person receiving it. Reread that email before you send it and pretend you're reading it for the first time. Does it make sense? Is the tone right? Is it better placed as a face-to-face conversation?

If you feel that your message belongs on a different medium or should be written differently, then it probably does. Adapt yourself for the benefit of others. Make that email a conversation, or that chat message a longer and more considered email. Your staff will notice the time and care that you put into communicating, and they'll respect that. They'll likely return the favor to you and their colleagues also.

Make Your Conversations Two-Way

Have you ever spoken to a senior staff member, only for that staff member to talk *at* you, rather than *to* you? Don't be that person. Conversations should be transactional between both parties. Ensure that you connect at a human level with the people that you are communicating with. Again, you're not a management robot issuing instructions. You're a human being connecting with another human being. Be natural and inquisitive. Listen. Ask questions. Invite responses.

Ensuring that those you speak to have air space in their conversations with you shows that you care about their opinion. It develops trust between you. It ensures that at the times when you definitely want your staff to tell you what they think, they'll be comfortable in doing so.

Respect the Preferences of Others

Be aware that different people have preferences over how information is delivered. This is an excellent subject for your one-to-ones. For example, some people may expect that the news of a salary increase be delivered face-to-face, whereas others may expect a formal letter or email. Depending on the person, getting this wrong could be jarring or offensive.

Dig into some example situations with your direct reports and note their preferences for different types of information. If you saw that they gave a good presentation, how would they like to receive praise? What about if they brought the production system down? You'll be surprised by how varied the answers can be from person to person.

Suitability Trumps Efficiency

In addition to being sensitive to the recipient when delivering information, be aware of your own bias to save yourself work when doing so. For example, it's extremely easy for you to send an email to your entire team rather than talk about the content of that email with individuals face-to-face. However, you should always choose the delivery that is best for them rather than you. It's more work, but it's worth it.

Be Consistent

Have you ever worked with someone who has been an excellent verbal communicator, only to receive an email from them later that was full of typos, bad grammar, and bizarre punctuation? How did it make you feel? Perhaps after an inspiring meeting with the company's new senior executive, you receive a follow-up instant message that could have come from an AOL chatroom in 1996:

```
hey mate u gt 10 min? soz 4 the dm lol
```

Perceived inconsistency between your role in the organization and the persona that you adopt in spoken, written, and nonverbal communication can feel really weird for others that you interact with. Try your best to unify how you communicate across different platforms. It makes others feel that they're talking with the real you, and each interaction is an opportunity to better your relationship and to build trust.

In addition to your consistency across mediums, is your formality also consistent? Sometimes chat applications can be highly informal, which can make it difficult to then interrupt the fun with critical feedback. You get to choose the persona you inhabit and the level of formality you bring with it. Fundamentally it's up to you, your team, and your workplace culture as to how formal you are in your interactions, but nobody wants a manager who is also the class clown. A useful way to approach this is to start formal and gradually ease the formality with time to a place you are comfortable with, that allows you to still be authoritative and critical when needed. Once you've slipped down to extreme informality, it can be hard to come back up again.

Remember that as a manager, your word has weight as it doesn't only come from you: it comes from an established position in the company's org chart. As a manager you represent yourself, your team, and your company as a whole. Be a role model for those around you.

Your Turn: Review Your Communication

Have a brief pause and do the following:

- Go back and read your sent emails, chat messages, and instant messages that you sent last week. Do you use a consistent voice across all of them, or do you change depending on the medium?
- How do you feel about the persona that you assume in the above mediums versus the persona that you project when you speak to people verbally?
- Do you work with anyone who comes across identically regardless of written or spoken communication on all platforms? Why is this?
- Do you work with anyone who comes across differently depending on how you're communicating with them? Why do you think this is?

How to Give Feedback

Entire books are written on giving effective feedback—and for good reason. Being able to deliver effective constructive criticism is a surefire way of making your staff improve and seek your counsel as a leader. What's even better is that they'll begin to do it to you, which will help you improve in turn.

The best and most concise advice on how to deliver feedback is the concept of *radical candor*, from the *book [Sco17]* of the same name. It outlines what to do and what not to do when talking to your staff. Learning the concept is simple, but mastering it requires practice.

When delivering feedback, you need to ensure that you:

- *Care personally* about the individual that you're talking to you. This ensures that you communicate with them because it's in their best interests and you want to see them improve.
- *Challenge directly* to ensure you say exactly what you think. Get to the point, and if you think that something isn't good enough, say it.

These two traits of sincerity and specificity ensure that you strengthen your relationship with your staff and push them toward growth. But what happens when you only do one and not the other, or don't do either of them at all? Have a look at the quadrant on page 49.

Looking at the diagram, you can see four different situations that can arise, depending on whether or not you care personally and challenge directly. Let's look at each of them.

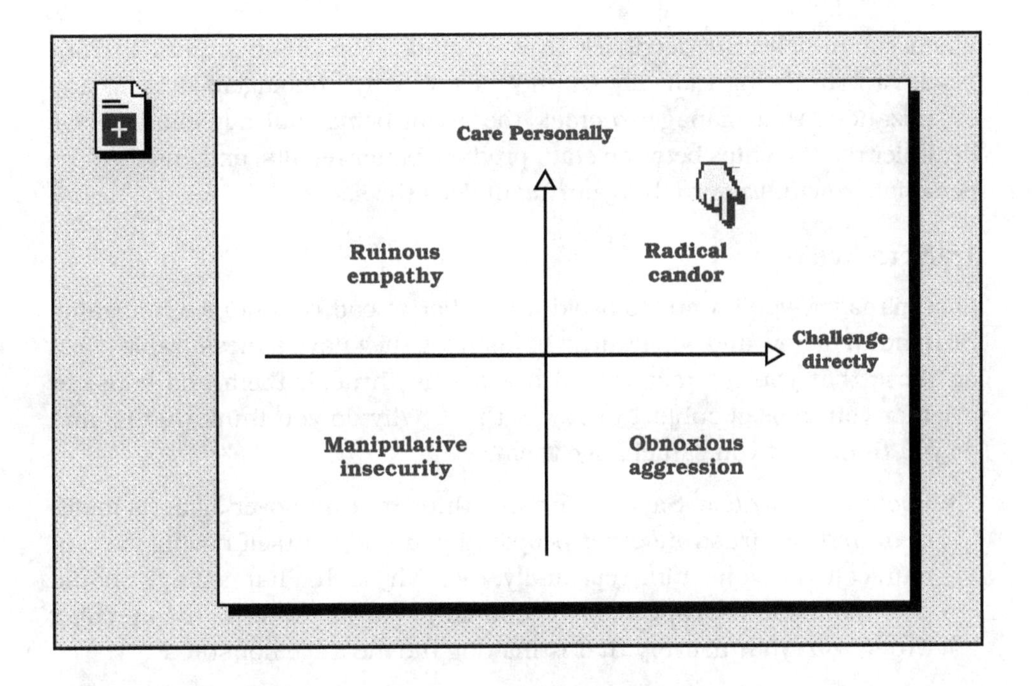

- *Radical candor* is what happens when you care personally and challenge directly. You say what you think and deliver the feedback precisely but in such a way that ensures the recipient sees that you are doing so because you care. This is where you want to be. It promotes trust and growth.
- *Ruinous empathy* happens when you care but don't challenge directly. It's either unspecific praise or diluted criticism to avoid a difficult message. This doesn't help the recipient improve.
- *Obnoxious aggression* is when you challenge but don't care. It's insincere praise or mean criticism.
- *Manipulative insincerity* is when you do neither. Good behaviors go unpraised, and bad behaviors are ignored.

Can you think of situations that you've witnessed in your own life that fall into the categories above? Perhaps you worked with a colleague who was performing badly and was a burden on the team. It may have been because your manager was demonstrating ruinous empathy by never tackling the performance problem directly. Conversely, you may have witnessed a manager who seemingly never praised their team and would get angry and overly personal in their criticism. Would you place them in the obnoxious aggression category?

When you're delivering feedback to your staff, keep this quadrant in your head. Are you being radically candid? If not, why? Managers, teams, and organizations that manage to crack the art of being radically candid have healthier relationships between staff, produce better results, and enable more personal growth. Ensure that your team does the same.

Traits to Avoid

As a manager, you'll want to avoid a number of bad communication traits. Be mindful of how and when these happen, as they have a measurable effect on those that you interact with. Have a look through them and consider whether you've been subject to any of them. Why do you think that is, and do you think that you can change them?

- *Overcommunication.* Saying the same thing over and over again is inefficient and can frustrate other people. If you find yourself having to communicate the same thing repeatedly, then why is that happening? Should that information be captured more formally in an email? Is there something wrong with your delivery that is making the message not stick?
- *Waffling.* We've all observed politicians in interviews avoiding questions by talking. And talking. And talking. Don't waffle. If you don't want to answer a question, or you don't know the answer, then just say so.
- *Playing to the crowd.* Nobody wants a manager who does whatever it takes to please them, since whatever it takes is typically hiding information, sugar coating it, or downright lying. Say it like it is, even if it's bad news.
- *Inconsistency.* Ensure that you're being the same person to everyone, and that your message remains consistent with everyone. If you're flip-flopping with decisions, then you've probably communicated them too early.
- *Letting emotion get in the way.* If you find yourself getting emotionally charged, angry, or frustrated, then be open about how you feel. Reschedule meetings if you need to. Just don't take it out on other people or let it cloud your judgment.

May the Communication Be with You

Wow, we've covered a lot! There are entire books written on communication, but this digest should serve you well as you get started on your management journey. Let's recap what we've learned in this section:

- We've explored the different *mediums* of communication—spoken, written, and nonverbal—and how you should use them to best suit the message you're delivering. You should be mindful of your mood and energy when going from activity to activity, and you should only communicate when you need to, not when you have the urge to.

- We've seen that communication *is not about you*. It's about the recipients. You should ensure that you listen, have two-way conversations, and respect people's preferences for interactions.
- *Consistency* is important for your own voice across spoken, written, and nonverbal communication. You should be aware of the weight your voice has as a manager and be a role model for those you interact with.
- The *radical candor* concept was introduced as a mental model to ensure that you're giving effective feedback to your staff.
- We've explored some *bad communication traits* you should try and avoid.

Delegation

Now that we've gotten our strategies for communicating well with our staff sorted, we'll look at one of the most powerful tools that you have as a manager: *delegation*. Remember that as an engineer, you write the code and delegate the work to computers all the time. Being an effective manager involves learning how to do this with people. We'll show you how.

Simply put, delegation is the transfer of responsibility for a task from yourself to a direct report. For example, if you've been asked to produce a report, then you can delegate that task to somebody on your team, and then when the work is finished, it can be given to the person that asked for it. All the while you can be doing something else. That's delegation. It sounds simple, right? Well, it is. It's not rocket science. However, a number of nuances with delegation make it difficult to master. But with a simple framework, you too can become an expert delegator.

Bad Delegation: The Two Extremes

To illustrate why delegation is important, let's look at two common symptoms of managers that are unable to delegate properly. You may have observed these behaviors in other managers you have worked with, or if you've been a manager before, then you may have even observed them in yourself.

The first common symptom is what happens when managers are *unable to delegate at all*. Have you ever observed a manager who always seems to be in a flurry of stress and busyness, with a whirlwind flying around them as they struggle to do just one of the 600 things they need to get done? All the while, their team is underworked or even idle. This is a symptom of not delegating enough and can happen for a few reasons. One is that in an industry where strong individual contributors often get promoted into management roles, they're unable to make the transition from doing all of the most important and critical work themselves to delegating it all to their team and giving them the mentorship and support they need in doing those tasks. I've worked with managers who are like this. As hard as they try, they inevitably do work to a poorer quality, subject themselves to a higher amount of stress and frustration, and ultimately their team suffers because they feel untrusted to do the important work. You don't want to be like this: that's why you need to learn how to delegate, and to delegate well.

The second common symptom is the inverse of the previous situation. The opposite of not delegating at all is to *fire and forget*. A manager gives a task to a member of their team and then completely forgets that it exists. This means that the manager is completely unaware of whether the task gets done or not, and to what quality. In the best case it may be done extremely well, but in the worst case it doesn't get done at all. What often happens is that it gets done not quite right, to a level of quality that isn't quite good enough, to the wrong deadline. This harms both the manager, who didn't get the work delivered, and the staff member, who didn't get the support and guidance that they needed to succeed.

If you've been working in industry for a while, you've probably seen countless examples of these two symptoms of poor delegation. It's bad for both the manager and their staff member. However, hope isn't lost. Learning the principle of delegation is actually very simple, and applying it isn't too hard either. Let's go through these principles together. They'll help you understand why these situations occur and prevent you from making the same mistakes. With time, you'll delegate masterfully without even needing to think about it.

You Do Not Delegate Accountability

The key to good delegation is understanding the difference between *accountability* and *responsibility*. Despite the words often being used interchangeably, there's a subtle but important difference between them when it comes to delegation:

- If you have *accountability* for a task, then you are being held to account for the completion of that task to the required level of quality.
- If you have *responsibility* for a task, then you're the one who is actually doing it.

We can define delegation by using these terms. Delegation is giving responsibility of a task to somebody else, while maintaining accountability for it. Sometimes it's easier to grasp this concept by imagining the chief executive officer (CEO) of a large multinational company with thousands of employees. The CEO is ultimately accountable for everything that the company does; however, it's impossible for CEOs to do everything themselves. No CEO is going to be a hands-on expert at finance, operations, sales, product, customer success, technology, and marketing. Therefore, the CEO employs leaders in those areas to which they delegate the running of those functions.

The CEO remains accountable for the technology of the company but delegates the responsibility for it to the chief technology officer (CTO). As we go further down the org chart, the CTO will delegate the running of various parts of their department to various vice presidents (VPs), who delegate to engineering managers, who delegate to their engineers. If one of the engineers was to maliciously leak sensitive customer data, then ultimately the CEO is accountable and will have to take corrective action accordingly. The engineer will not be the one answering the questions from the press.

Looking at the two examples of bad delegation in the previous section, we can now reason about what is wrong with them. In our first example where the manager is overwhelmed from not delegating, it's because they are assuming accountability and responsibility for all of their tasks. The fix is to give responsibility to others and delegate. For the second situation, the manager that is firing and forgetting is giving responsibility of the task to others but is also giving away the accountability, which is wrong. This is the opposite of delegation: abdication.

The Scale of Delegation

Successful delegation comes from giving the responsibility of a task to somebody else while maintaining accountability for it. There's an art to doing it well. We can use a scale to look at how a task should be delegated to an individual as shown in the image on page 54.

As you go from left to right on the diagram, the amount of delegation is increasing. It's important to note that full delegation can only happen when the person being delegated to can perform that task exactly to the standard that you expect with no guidance. The scale between no delegation and full delegation involves applying the right amount of teaching, coaching, mentorship, and check-ins. That is why retaining accountability is so important: not only does it ensure that the task gets done properly, it also gives your staff a chance to learn and grow.

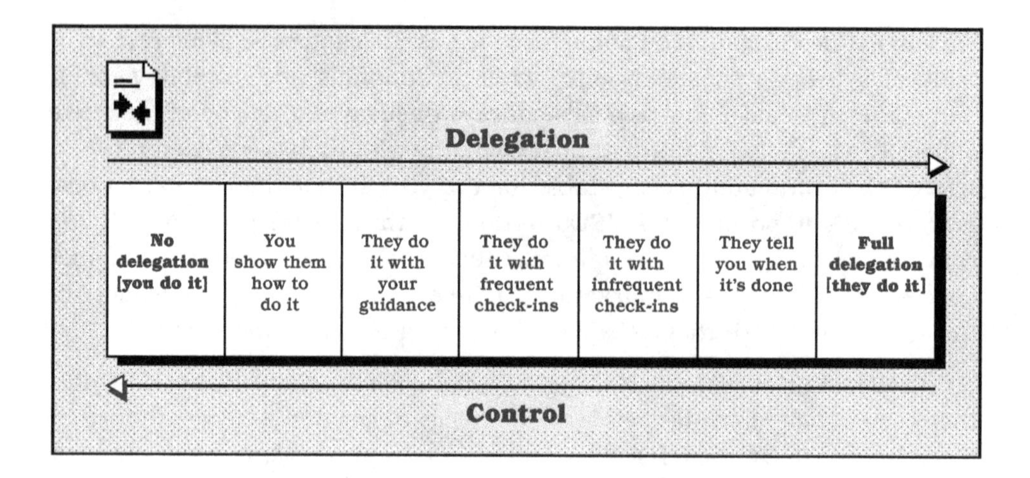

It's important to remember each *task* is delegated individually. This means different tasks should be delegated in different ways. The reason for this is people's expertise differs between tasks. It isn't correct to think a junior engineer will always need careful coaching through everything and a senior engineer only needs to be told what to do. Nobody is an expert at every task, regardless of their seniority. As manager, it's up to you to judge each task and delegate it appropriately according to the place you assign it on the scale.

Another arrow on the diagram is moving in the opposite direction: control. With more delegation comes a loss of control, which in most cases is absolutely fine. But you may find situations in which you delegate less to maintain more control. For example, a critical task may shift the amount of delegation that you do to the left because of the nature of the work. For instance, if the production environment has been broken by the latest deploy, you may check in more regularly with your senior engineer that is fixing it than you would for the same bug fix on the development environment.

Here's the rule you need to follow: given the right mix of control you need for the task, delegate as much as you can. With time, your staff will learn more, and you can delegate more, increasing the output and effectiveness of your team.

Delegation Do's and Don'ts

Before we wrap up, let's have a look at some do's and don'ts of delegation.

Do:

- *Delegate.* You cannot be an effective manager without delegating your work. A team of people can produce more work than one individual can. Allow them to do so.

- *Delegate enough to challenge your staff.* Delegation is a win-win situation, since your team can get more done and your staff get to feel challenged, learn new skills, and accomplish things together.
- *Retain enough control depending on the criticality of the task.* For important, highly visible work, don't be afraid to move to the left on the scale to ensure that you're on top of what's going on.

Don't:

- *Abdicate.* You cannot delegate without maintaining accountability for the task. If you are not accountable for the result, then you are abdicating from it.
- *Expect others to work the same way as you.* The more you delegate, the more likely it is that the person will approach the task differently to you. That's absolutely fine. You should maintain accountability over outputs, not the process. Doing otherwise is not delegating: it's micromanaging and meddling.
- *Take things back.* You should be delegating tasks to the correct point on the scale so that your staff either do it themselves or learn how to do it. If you've chosen a point on the scale that is too far to the right, then slowly move it back toward the left. Never get frustrated and do it yourself. It's your responsibility to make the task as much of a learning opportunity as possible. Snatching it back and doing it yourself not only cheats your staff out of an opportunity to learn, but it makes them feel bad too.

Your Turn: Delegation

With the scale of delegation in mind, where on the scale would you place the following tasks?

- A medium-priority bug fix being done by a senior engineer in a system that they wrote themselves.
- A new feature being written by a junior developer.
- An intern setting up their development environment for the first time.
- A principal engineer working on a new distributed storage architecture.
- Another team lead who is your peer whom you've asked to implement an API endpoint for your team to use. They've said their team will do it soon.
- The CEO, who has said she will get board approval for the $250,000 cloud investment for your new product infrastructure.

Delegating the Implementation to You

That's it—delegation is simple, right? It's easy to learn, but tricky to master. But with continual practice, you'll be an expert delegator if you follow the principles we've outlined.

Working with Your Manager

In this chapter we've looked at the skills you'll need to work with individuals: strong communication and the ability to delegate. However, there's one important individual that you probably didn't have in your mind while reading. That individual is your manager.

For many, their manager is the person that tells them what to do and judges their performance. But the relationship can be so much more if you focus on developing it by using the techniques that you'll learn in this section. You can make your time with your manager a part of the week that you value and that you look forward to.

Let's have a look at some of the techniques that you can use.

Pull, Don't Push

The most important piece of advice is that you should *pull* on your manager, not wait for them to *push* to you. What this means is that it's up to *you* to get the best out of the relationship that you have with your manager. Be proactive and don't wait for them. Book that meeting, have that conversation with them, ask for that help. You're not a marionette dangling from the strings that your manager holds, with them dictating exactly what you should do. Instead, you need to reframe the relationship that you have in your mind as one that encourages two-way empowerment through coaching, sharing of responsibilities, and support.

The techniques that follow all embody this principle. You need to understand how they view you and your team within the rest of the department and work out how best you can make it a better place. You need to remember, like you, your manager is probably busy and has a lot on their mind. They may be in a situation where they're not getting the support they need from their own manager, so you can form a strong partnership by being on their side, with both of you supporting each other and getting meaningful work done.

Talk About Performance

At the beginning of your relationship with your manager, you should talk about performance. Notably, there are two areas to explore:

- What does *your performance* equate to in the eyes of your manager? Is it the retention and happiness of your staff, the amount of work that you get done on time, some strict uptime SLAs, or is it entirely subjective?

- What is *their performance* dependent upon? You can begin to get a view into the wider organization. Are they fundamentally accountable for everything that their teams ship, or are they accountable for more abstract business metrics? Do they even know?

You may even find that the answers are not clear to either of these questions. If they aren't, then this is an excellent chance to define them. Perhaps your manager feels that good performance means that they need not worry about the internal workings of your team at all, and instead they only need focus on their relationship with you. Alternatively they could be accountable for ensuring that the department ships three major projects this year, that they retain 90% of their staff, and that a specific SLA is upheld.

If you've been a programmer before, then you may have thought about your performance and the performance of your colleagues. If you had to rate the performance of another programmer, then you may have considered:

- What level of technical expertise does this individual have? Are they able to take on large, ambiguous, and technically challenging problems and solve them in code? Or do they need others to help break down the problems so that they can implement each individual piece in isolation?

- How much do they increase the skill of the whole team? Are they a patient and skilled mentor who takes the time to teach others through conversation, demonstration, and collaboration?

- Do they have a positive effect on the team? Are they motivating and optimistic, and do they lead by example? Do others follow what they do?

- Do they understand the bigger picture of the software that they are building and the needs of the company?

I would advise that you work on framing your own role in the same manner with your manager. It's also a good exercise for them to frame their role in the same way so you can see how your team fits into the bigger picture. Sometimes it can be difficult to define managerial performance because it's more abstract and your output—as highlighted before—is the collective output of your team and those that you influence.

Later in *Dual Ladders* we'll be looking at defining career tracks for both individual contributors and managers. Feel free to skip ahead to that chapter if you'd like to have more material on hand for trying to frame how your performance is measured. However, if you'd like some inspiration for an initial conversation with your manager, then you could incorporate the following ideas:

- What *traits* does a high-performing manager have? What do they say and do regularly that make them so?
- How do you *measure* that a manager's team is doing a good job? Is it simply through the output of their team, or is it through team happiness, retention, or all of these things?
- In the spaces between doing hands-on managerial activities, what is a successful manager *doing*? Are they getting stuck-in with coding, or are they working with their manager and peers on making the department a better place? If it's the latter, then what sort of activities does that entail?
- How *connected* should a team's manager be across the organization? Should they primarily be concerned about their team, or should they also be regularly talking to designers, product managers, and sales and customer success staff?

Getting both your input and your manager's input on these themes—plus any others that you can think of—helps frame how you both think about performance. It also ensures that you talk about it from the start, which is a key part of a radically candid relationship.

Lifting the Lid on Their World

As you get to know your manager, you'll want to begin to lift the lid on their world. If you wait for your manager to tell you what to do, or ask you for your input, you may be missing out on learning opportunities and career growth.

A simple technique is to ask questions during your one-to-one meetings. This has a number of benefits:

- It exposes you to the concerns that are going on at one level up in the org chart.
- It can expose additional things you can help your manager with. For example, they may mention another team is having difficulties with a particular project or technology, and you may be able to offer some advice.
- It shows that you're interested in career progression and growing outside of the confines of your role.
- It strengthens your relationship because you are taking an active interest in their world.

The questions themselves can be straightforward and worked into your conversation:

- "So what's been on your mind this week?"
- "What's your biggest worry at the moment?"
- "How are your other staff doing?"
- "What are your peers working on?"

Although simple, having this active interest can unlock a number of career growth opportunities for you. You may get to assist with hiring in other teams, expanding your influence. You might get a chance to meet others in the organization that you can form a relationship with. Maybe you'll even get asked to assist with performing technical due diligence on a startup that your company is thinking of acquiring. You never know. But it starts with questions and an active interest. With time, you may see your relationship with your manager feeling more like a meeting of peers rather than a boss and their subordinate. You both get real value from each other.

The Power of Summaries

In the same way that you can ask questions to get an insight into your manager's world, there's also a way that can open up yours to your manager while also helping you pause, reflect, and plan. That technique is the simple regular written summary. Every week, take thirty minutes to note down the main activities that you and your team have been involved in, how you feel about them, and what decisions you've taken or need to take. Write down how your staff are getting on and how they are feeling about their work. Is anything becoming a blocker? Do you see trouble ahead? Is everything actually just going really well? Note it down.

The simple act of taking time for yourself every week to summarize how things are can be extremely therapeutic. It can help you abstract away from the problems that you are facing and work them through with yourself on paper. Often you'll find that they're not so important or challenging once you've spent some time with them. Studies have shown that journaling in one's personal life can have a positive effect on mood and well-being. I've found the same to be true about weekly journaling at work.

If you struggle with what to write, then perhaps you could try an approach based on four Ps:

- *Progress*: What has happened since last time you wrote?
- *Problems*: What issues have occurred and what needs addressing?
- *Plans*: How are you going to approach those issues going forward?

- *People*: How are the individuals on the team? Are they doing well or are they having a hard time? What could be improved?

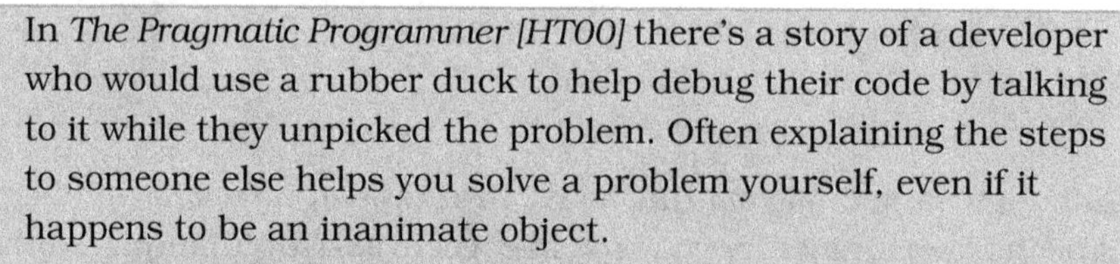

Rubber Ducking

In *The Pragmatic Programmer [HT00]* there's a story of a developer who would use a rubber duck to help debug their code by talking to it while they unpicked the problem. Often explaining the steps to someone else helps you solve a problem yourself, even if it happens to be an inanimate object.

Writing weekly summaries is the closest thing that I have found to rubber ducking in a managerial role. Often when I've had multiple issues whirring around my brain, the act of committing them to paper has enabled me to be more logical and clearer in my thinking, thus unlocking the solution. Sending them to my manager weekly holds me accountable, so I always perform this summarization exercise. Better still, it gives us a lot to talk about.

Here's what you can do to build these summaries into your week. Before you start doing this with your manager, let them know that you're going to do it. You can tell them that you don't necessarily need a reply, but you're interested in showing them weekly what is going on in your world. They are, of course, totally welcome to comment, debate, or challenge anything that you write. In fact, that makes excellent material for your one-to-ones.

- *Pick a suitable time* to do it and make it a recurring to-do. Perhaps you find that finishing off Friday afternoon by writing is a good place to pause and reflect on the week that has just ended. Perhaps it's more useful to do it the day before you have your one-to-one with your manager so you can talk about the issues in greater depth when you next meet.

- *Dedicate thirty uninterrupted minutes* to writing. Make some high-level bullet points on the areas that you've been focusing on, whether those are elements of the project, interactions that you've had with your staff or outside the team, concerns you have about what's coming next, or any progress toward your goals.

- *Write as you think* in a continual flow. Don't worry about formality. Write as if you were speaking to a good friend. How have you felt about each item on your list? Have you worked through any problems, or are you stuck on something in particular? You may find that you make progress on your problems as you write.

- *Give it a quick proofread.* No more than a couple of minutes. Don't worry about it being perfect.
- *Send it to your manager.* I recommend using a shared document that allows commenting. That way, you and your manager can continue the discussion asynchronously.

This technique is especially useful if you have a remote—or extremely busy—manager and don't have the opportunity to walk over to their desk each day to have a conversation. However, I have used this technique identically when I've had my manager sitting on the desk next to me and when they've been on the other side of the Atlantic Ocean. The ability to perform an exercise that benefits both you and them is a high-leverage activity and it comes fully recommended.

Your Turn: Work Closer with Your Manager

That wraps up a number of techniques that you can use for working with your manager. Try the following and see how they go:

- Identify situations in your week where you feel that you should wait for your manager to interact with you. Next time, be proactive and go and interact with them first. Run a thought by them or ask their opinion on something. How does it feel to be doing this rather than waiting for them?
- In the next chapter we'll be covering one-to-ones. When you sit down with your manager, bring up your performance and their performance and how they are measured. What did you learn? Are there clear indicators of good performance or is it entirely subjective? How is *their* manager's performance judged?
- Make an effort in the coming weeks to lift the lid on your manager's world. What are they working on? What's on their mind? What did you discover that was interesting to you?
- Schedule in a weekly summary to your manager. Begin the habit by starting on it this week. How did you feel doing it? Did it help you pause and reflect on what you've been doing?

Onward!

Well done on making it through a lot of material on working with individuals! We humans are challenging, aren't we? Here's what you've learned in this chapter:

- How to *communicate well* through spoken, written, and nonverbal communication.

- A framework for *delegation*, which may be one of the most important things that you learn as a new manager. Many managers either don't delegate properly or do it badly, so you can have a real head start if you begin practicing this.
- How to *define the relationship with your own manager* by pulling on them rather than waiting for them to push to you, keeping your performance as part of regular conversation, and encouraging them to share more about what they're doing with you so you can build rapport, increase your knowledge, and contribute at a higher level.

You'll see that with time, all of these concepts will become second nature. However, some, like delegation, are easy to understand yet extremely tricky to master. In the next chapter you're going to take everything that you've learned so far and jump right in to scheduling, planning, and performing your one-to-ones. We'll make you a pro in no time.

Give me a lever long enough and a fulcrum on which to place it, and I shall move the world.

> *Archimedes*

CHAPTER 4

One-to-Ones

This afternoon you begin your one-to-ones with your team. You've got three of them booked in today: Ben, Katie, and Lee. Your initial meetings with them last week went swimmingly, but you were mostly just gathering information about their roles within the team and the projects that they are working on. Today it's down to business, and you're feeling the stage fright.

To the content: you had a bunch of ideas last night that felt inspirational, motivating, and life-affirming. But on reflection this morning, they seem awkward, cheesy, and dumb. You nervously tap your fingers on your knee. You want to get these relationships off to the right start. You worry you're going to be spending several hours sitting in painful on-off silences. What if you don't have anything interesting to say? And what are you even *meant* to be saying? Your previous manager was broadly absent, so you don't have experiences from the other side of the table to model. You begin to get a bit worried.

Taking out your phone, you Google "what to talk about in one-to-ones." You scan the results from the many listicles that populate the first page of hits. "Self-critique"? It's a bit early to dive into that. "Check in on objectives"? Well, you haven't really set any with anyone yet. "Talk about pay"? Erm, that doesn't seem like a normal first meeting. "Coaching"? But on what? You still need to get to know these folks first before you can get into a rhythm of meaningful conversation.

You hop off the subway at your downtown stop. Ascending the station stairs and emerging onto the street, you hope for an epiphany, otherwise it's going to be an extremely awkward afternoon. As the traffic noise of the city swirls, you can't help but tune in to a loud conversation happening between a pair of suited men walking in front of you.

"So did they get the contract?"

"Yeah. We had to make some amends and also walk back on some of the pricing, but it's signed."

"Nice job. What were some of the changes?"

"Nothing major. They want some regular support from our services team, which isn't a big deal. They also want a QBR, plus the terms to be re-evaluated at the end of the year."

You reach the crosswalk and wait for the traffic to stop. If only you could have a contract with your direct reports about what each side needs from the relationship, you think. The light turns green. You step out into the road. That'd make things easier at the start. Hang on a second. This sounds like a good idea...

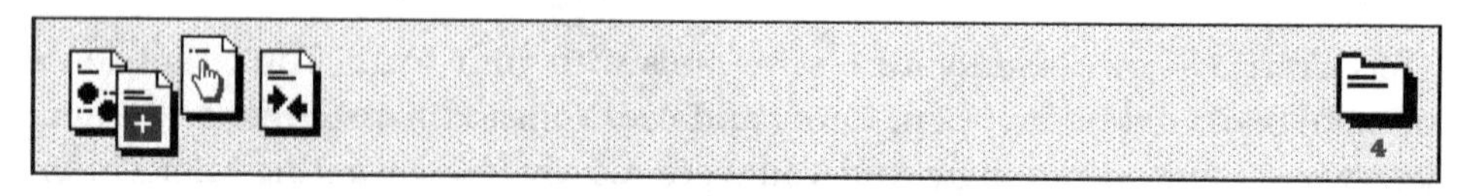

In this chapter, you're going to learn how to do one-to-one meetings with your staff. Although you may be wondering how this subject can generate an entire chapter of content, I can assure you of the following:

- That you're going to be spending a lot of time in one-to-ones, so it's in your best interest to be able to do them confidently, comfortably, and to get the most out of them.
- That there is a vast difference between having an unprepared chat and having a meaningful, focused one-to-one.

So that's what this chapter is here for: to give you the context, techniques, and inspiration that you need to have great one-to-ones.

In this chapter you're going to learn the following:

- How to best *prepare* for one-to-ones, again touching on why they're an essential part of your role.
- How to do your first one-to-ones by using an exercise called *contracting* in order to get your relationship off on the right foot.
- Ideas for *what to talk about* each week, including how much you should actively participate versus passively listening.
- How to *take notes and assign actions*. I'll show you the system that I use to do it so that you don't forget anything.

OK, are you ready? Let's get going.

Week In, Week Out

One-to-one meetings form the core of the relationship you have with your staff. They are the point during each week in which you both give up your time to be together. High-quality one-to-one meetings build high-quality relationships. They allow you to engage with your staff on a deep, personal level. By performing them mindfully and to a high standard, you can have a tremendous impact on the performance of your staff, and even on their lives. If you manage people for long enough, your one-to-ones will cover pretty much everything, from good times to bad, to conversations with highly performing staff to those who need to leave the company, to news of births and deaths, both of real people and of JavaScript frameworks. They really are life's rich pageant.

Let me use a buzzword for a second: *leverage*. When we talk about leverage in a business context, we often mean doing highly impactful activities where a small investment of time and effort can result in large amounts of output. If you look back at the epigraph on the first page of this chapter, you'll see it's a quote from the Greek mathematician Archimedes, and indeed, with a large enough lever you *could* move the world. As a manager, you'll be looking to find those opportunities to exert large amounts of leverage, and you're in luck: your weekly one-to-ones with your staff are one of the best places for that to happen.

You'll want to do one-to-ones with each of your staff every week. Yes, every week, without fail. A steady cadence and regularity are key for these meetings. It brings predictability into your relationship, demonstrating that you're there for your staff week in, week out.

But first, let's get prepared.

Leverage

Seeing as you work through other people as a manager, you should be keeping the Archimedes quote at the beginning of the chapter in mind. Where are situations where a small amount of your effort can have a disproportionately large outcome? These are the areas in which you should be focusing.

For example, that difficult conversation, once had, could completely change the direction of an individual's performance, making the team happier rather than others having to pick up the slack. Additionally, if you can master delegation, then you can have a whole team effectively doing your work, which is greater leverage than trying to code faster and more efficiently yourself.

Leverage

> Think of situations where you can use the idea of leverage to best spend your time creating the biggest possible output. Where are they? Why are you not doing that activity already? Is there a difficult conversation that you're putting off?

How to Prepare for One-to-Ones

As you've been applying what you've been learning in the book so far, you'll already have a recurring weekly meeting booked in with your direct reports. If you haven't done that yet, it's time to do it.

Ideally, you'll want them to be at the same time and in the same place each week, using video conferencing software if one or more parties are remote. Don't move them around unless absolutely necessary: you'll want to have a predictable, safe, regular touchpoint with your staff that they can rely on. With time, both you and your staff will batch up things to talk about as they unfold over the course of each week.

Make sure that you're having the meeting in a private area, such as a meeting room. Don't do them in a public area of the office, such as the breakout area or near their desks. Doing so risks any private conversation occurring being overheard and may also carry an implicit feeling that you would rather not hear private matters at all, due to the location. You want to be able to have candid and transparent conversations, since that's the valuable and interesting content, so ensure that it's able to happen. As much as I like going to a coffee shop, a private room is less noisy, less likely to accidentally have other staff overhear sensitive information, and allows both of you to really focus on each other. After all, given that one-to-one meetings are about personal development, you owe it to each other to find a quiet space rather than having to shout over a hissing milk wand.

The second piece of preparation you'll need to implement is a rolling agenda that guides your meetings. The easiest way that I've found to do this is by using a shareable document, such as Google Docs. I have a shared private document with each of my direct reports that contains our agenda. A private document is a great way of recording what is discussed within your one-to-ones and capturing actions that either you or your staff have to do. I've gotten into the habit of jotting things in this document over the week as situations unfold and thoughts arise. So have most of my staff. That way there's always plenty of material to talk about on the day.

So, let's recap the preparation that you need to do. Have you:

- Booked in a weekly one-to-one with each member of your staff in a private space where you won't be interrupted?
- Created and shared a document that allows you to both keep track of your agenda?

Have You Booked In Your One-to-One with Your Manager?

Before we move on, have you set up your one-to-one with your manager yet? If not, then get that done. Do the exact same thing as with your staff: a weekly private meeting and a shared document for notes.

Although you may think it's best you wait for them, we showed in *Interfacing with Humans* how important it is to be proactive with your manager to get you the support you want. They'll appreciate it.

Contracting: Your First One-to-Ones

Your first one-to-one meeting can be somewhat harder than the ones that follow. Just like meeting any human being for the first time, your first one-to-one could unfold in a multitude of different ways. You might be lucky enough to immediately gel with your direct report: the conversation flows and away you go. Equally, you might be unlucky enough to sit down, stare at each other, and not really know what to say. Depending on your direct report's experience in their career, the company, their cultural background, and their personality, they may approach this meeting in different ways. Could there be a method in which to normalize this first meeting so that it works better for both parties?

If you've started your managerial role in a new organization, then you will not know your direct reports intimately, which may make it challenging to judge the right level of formality to bring into the meeting. On one hand you're their manager, but on the other you need to be their close ally. You don't want to be too serious as you may risk causing alienation, but equally, being too jovial might come across as strange depending on the person and the existing culture of the organization.

Even if you've been promoted from within, that doesn't necessarily mean that these meetings are going to be any easier. In fact, they could be more awkward: your relationship could be changing between you and your colleagues. Close peers could now be subordinates, and that's a strange and potentially difficult change to navigate through, especially if you and your peers have become good friends. This is starting to get complicated isn't it? What should you do?

Rather than risking your new relationship with your staff getting off to the wrong start, there's a useful exercise that you can both follow that allows both parties to openly talk about what they expect from one another and then outline their wants and needs from the relationship. This exercise is called *contracting*, and you're going to learn how to do it.

What Is Contracting?

Contracting is simply a set of questions that you can work through in your first one-to-one that provokes conversation about what is expected from you and your direct reports. These expectations come from both sides of the relationship. The contracting exercise forms a wrapper in which you can talk about each other's needs in a structured and safe way, promoting candid and transparent conversation that should continue into the future.

When preparing for your first one-to-one, explain that you're going to do a short exercise to understand how best you can support your direct report as their manager. How you prepare is up to you to decide: you could leave it until the meeting to reveal the questions, or you could forward them to each of your direct reports ahead of time so that they can consider them carefully. With time, I've gravitated toward the latter, since the quality of content tends to be higher.

Remember that *both of you* should be answering each of these questions in relation to one another. This isn't a one-way exercise. You support each other in your relationship. With the questions at hand, spend your first one-to-one working through them with each other. With a bit of luck, you'll find that:

- You both will have been able to discuss what you hope, want, and need openly in a structured way that does not make either party feel uncomfortable.
- You'll have begun your relationship by talking candidly, which should set the tone for further interactions.
- You'll have plenty of material to talk about going forward.

Let's have a look at each of the questions. They're the same ones that I use in my own contracting sessions. But they should only serve as a guide. If you want to change them to suit you, then by all means do so!

1. Which Areas Would You Like the Most Support With?

The first question is a broad one. It's broad on purpose: each of you may need support in potentially any area. For example, you as a manager may need help from your senior engineer in better understanding the technical challenges that the team faces in their technology stack, and they may need more help in promoting their own architectural ideas within the team and beyond. As well as technical challenges, there may be interpersonal issues that require support, such as being able to better work with difficult colleagues in their team or working on self-confidence in debate and discussion.

Resist the urge to suggest areas yourself if they are struggling to come up with them. Seeing as you are their manager, they may assume you're suggesting them for some purpose and agree with you, when in reality all they needed was some more time for those answers to unfold. Simply say it's fine not to have clear ideas right now, but you're there for them at any time they need you. Also, solutions are not required just yet. You can turn all the talking points into more lengthy conversations in further one-to-ones, thus building a thread of continuity throughout your meetings. If your direct report struggles with building consensus with their technical decisions, you can come back to this particular aspect of their work continually in the future, with reference to actual events and situations they've been in.

Make sure that you note everything down in your shared document as you talk. Now it's on to the next question.

2. How Would You Like to Receive Feedback and Support?

This is about working out how the other person likes to operate. Everyone's personality is different, so it's important that you both feel comfortable with the manner in which you interact.

You may notice a wide variety of responses here, some of which highlight your own comforts and discomforts. If their answer is that they don't mind, then I'd advise probing a little deeper. Try and frame the question with examples. Let them imagine, for example, that they were in a meeting with the team, proposing a technical direction in front of the room on a whiteboard. Would they be more comfortable with you calling them out on an inconsistency in their approach in front of the whole room, or would they prefer you told them privately face-to-face afterward? Do they actually prefer receiving technical feedback in writing so that they can digest it better? Once you begin to lay out some examples, you may find that your staff are all very different.

It's important that you find out the best way in which to deliver your feedback and support so that you can have the greatest impact when doing so. Since you'll want to have a relationship where direct and honest feedback is given at all times, you'll want to operate in the way that allows the feedback to land in the most impactful and humane way possible.

3. What Could Be a Challenge of Us Working Together?

Like the first question, it's good to give them the airtime to think properly about this. They may have some first impressions of you that frame their interactions; for example, they may be less confident at speaking in front of large groups and therefore be afraid of challenging your opinion in those situations. Alternatively, your background could be in JavaScript and theirs in Java, and they feel that you might not be able to offer much support in terms of their technical development. All of these concerns are valid and now is your opportunity to discuss them in detail and consider some strategies. In the last example, one suggestion could be to delegate technical mentorship to a senior engineer with the same skill set.

As mentioned previously, your promotion into a managerial role may change the existing relationships that you have with staff at the company. How exactly are you and a colleague going to stay good friends now that you are your colleague's boss? Did somebody else in your team also apply for the role but not get the job? How do you both feel about this? Talk it through and think of some ways that you can both work through it.

4. How Might We Know if the Support I'm Offering Isn't Going Well?

If the relationship is taking a turn for the worse, then it's good for both of you to be aware of the signs. It could degrade in numerous ways. You could be frequently experiencing negative emotions from your interactions, or they could not be exhibiting benefits from your particular style of feedback.

If the situation was getting worse due to conflict, then this is an opportunity to explore how both of you react to a variety of negative emotions such as disappointment, frustration, and anger. One person may become furiously vocal in a conflict situation, whereas another may say very little but the issue is boiling inside them. You should be aware of each other's signs so you can spot them.

If your coaching and support are not working out, then it may be as simple as agreeing that it's OK to openly tell each other, and then further steps can be taken to improve the situation or for them to receive mentorship on particular skills from elsewhere in the organization.

5. How Confidential Is the Content of Our Meetings?

A number of sensitive issues may be raised during your meetings. For example, your direct report may say that their colleague's performance has been getting worse over the quarter. However, how confidential do they consider this information? Would they feel uncomfortable if you went and followed up directly about it? Would they feel uncomfortable if they mentioned that the feedback came from them?

Additionally, it's worth asking how they feel about the default confidentiality of the meetings with your own boss. Are they comfortable with you relaying anything that they say upward, or should you check before? Running through some of these scenarios can help tease out the answers if they are unsure.

Contracting: Examples from My Staff

Here are some real examples from contracting sessions that I've had with my staff so you can get a flavor for some of the aspects that can arise.

- I had a situation where a peer became a direct report. We discussed how we can operate so that it didn't feel like a demotion for them, mostly by ensuring that they maintained a lot of the decision-making autonomy that they already had.
- Another member of staff that moved into reporting to me had preconceptions of how I was as a person because I was quite outspoken. This wasn't entirely true, and we made sure that if I ever was that way, that they were perfectly cool to call me out on it and I wouldn't be offended.
- One member of staff was really interested in raising their visibility and profile in the department and thought I had done a good job of that myself. We worked that into their projects going forward, using similar methods that I'd used over time.

Contracting: A Summary

The contracting exercise is an excellent way to break the ice in your first one-to-one meetings. It increases your understanding of each other and sets some expectations between you and your direct reports. Go ahead and do it. I've always been surprised at the quality of the content that it generates. Contracting doesn't only have to be done once. It can be useful to revisit the exercise over time as both of your needs adapt and evolve.

If you get all of your contracting sessions done, you should be in a strong position with plenty to talk about in the coming weeks. That's what we're going to talk about next: how should your regular one-to-ones unfold, and how can you make sure that they are successful?

What to Talk About and How to Do It

With the initial one-to-ones handled by the contracting exercise, you have a clear structure for those initial meetings. But what about the next meeting and every meeting after that, forever more?

I tend to use some simple repeating patterns for content. First off, it's essential that you prepare. I have a recurring to-do list item that repeats on the day of each direct report's one-to-one reminding me to spend some time before the meeting putting the agenda in order. If I haven't already noted down some items to talk about, then I'll have a think and pre-fill the agenda with some items. These can be anything from:

- *Observations from the past week*, either about their work or the team's. These can be good or bad observations, or even just areas where I'd like to probe a bit further and find out more.
- *A deep dive* into a project or piece of architecture that they're working on.
- *Updates* that will be interesting for my staff, such as what has recently gone on in any other meetings that I'm part of, or in my case, anything my own manager has told me, such as relevant things discussed at the last board meeting.
- *Coaching*. We'll learn more about coaching in *How to Win Friends and Influence People*. You can help your staff pick through their own problems.

But even if you have fantastic content to cover every week, you can still get your one-to-ones very wrong if you don't approach them in the correct way.

It's Their Meeting, Not Yours

Despite the fact that your one-to-ones are your best chance to positively impact your staff each week, the paradoxical stance that you must adopt is that the meeting is theirs, not yours. What this means is that you must "keep the thought bubble over their head" for as much of the meeting as possible. Do this by asking leading questions, nudging the conversation in particular directions, and most of all, by listening. It's not your job to direct the conversation or pontificate in this meeting; it's your job to absorb and guide.

Try and get your direct reports to do 70% of the talking. If you feel like solving their problem for them, don't. Ask another question and let them arrive at the conclusion themselves. This is an art that takes some practice, although some are naturally good at this with little to no training.

You can achieve this by asking lots of leading questions, rather than providing all of the answers. For example:

- "How has your work been going this week?"
- "What are you working on today?"
- "Tell me about those production issues we had last week."
- "Do you think that we're measuring our uptime well enough?"
- "How are you feeling about our deadline in June?"
- "How best could we ensure that we've got all of the right metrics being logged ahead of time?"

Although seemingly convoluted, leading questions work well. They encourage good discussion, and they can be used whether you actually know the answer or not! Use them to your advantage.

Silence Is Golden

Don't feel like you've got to fill every moment of airtime with conversation. I've found that time and time again if you let the dialog unfold and stay relatively quiet using subtle prompts, the best parts of the conversation are to be found. For example, if your direct report is discussing some issues that happened in the previous week, letting the conversation tail off without replying can help surface an issue: "...and that's the problem, I guess. I just have no idea why they aren't helping us more."

Once again, this is why this is their meeting and not yours: let them dig into their mind and surface the issues that really matter to them. They'll know.

Updates: The Boring Part

One-to-ones are not status update meetings. Don't make iterating through the tasks that your staff are working on the primary purpose. Ideally there's already another place where you could find out this information if you wanted to, such as the team's tickets. However, there will always be *some* element of updates in these meetings from both sides.

But updates should not be the core element of the meeting. They're an aside. I've been in one-to-one meetings in the past where my manager has walked through a checklist of the tasks that are in flight and asked me for updates on each of them. This doesn't progress your relationship, fails to focus on personal development, and most of all, is extremely boring.

You can make updates more interesting by being creative. Instead of just nodding and listening to what they've been doing, why not probe deeper by asking some questions?

- "How could we deploy that into production quicker?"
- "Is this the correct technical approach? What are some alternatives?"
- "Have you seen any open source software that could solve that problem for us?"

You may have absolutely no idea of the answers to any of these questions, but they work well for stimulating discussion. Keep updates to the bare minimum required, though. Free up most of the space to talk about wider issues around the work and their personal development. You'll both be happier for it.

Pulling the Andon Cord

Particular directions that the conversation can go in your one-to-ones may make the other person feel uncomfortable. This can happen for a multitude of reasons: either a subject is too personal, too off topic, or requires the support of someone that is better placed to address it.

In his book *Toyota Kata*, Mike Rother details how production lines at the automobile manufacturer would have multiple cords that should be pulled by any worker as soon as they noticed a defect. This was called the Andon Cord: the word *andon* means sign or signal. It would stop the whole production operation so that leaders could solve the problem, and then production would resume.

You could implement your own metaphorical Andon Cord in your one-to-ones. If either party is uncomfortable with the current topic, you can stop the conversation and enquire as to why. Once that's resolved, you can continue.

Ideas for Topics of Conversation

If you have a particularly quiet direct report, or if you're just looking to mix up the material a bit, then here's a whole bunch of example topics that you're welcome to use for yourself.

- *Architecture deep dives*: At irregular intervals, I find it fun to ask my direct report to take me through the latest architecture of some part of the system that we're working on. I ask questions about various parts in terms of speed, resilience, redundancy, and so on. This sometimes highlights some weak points, but mostly I get a chance to better understand the work and the part of the infrastructure that it belongs to.

- *Process deep dives*: How many steps does it take to get something done, such as releasing code to production? Why is that? Could those steps be reduced, and if so, how? Could any processes be removed completely?
- *A relevant article you've seen*: I typically read a bunch of technology sites, such as Hacker News. If there's something interesting I've seen—either as a subject or as an open source project—I'll talk about it. We can discuss what we're doing in that area, if anything, and try to pick it apart. You'll find these discussions always find their way back to what you're working on.
- *Teaching*: Once you've finished this book, you'll have a whole host of topics to talk through with your direct reports. If you share them, then they'll be able to see more of the world through your eyes. For example, have they ever thought about delegation and how they could factor that into how they share work with less senior staff so they can learn new skills?
- *The department or company direction*: What do they think? Do they feel confident in where the company is going, or do they have reservations? If so, why is that? Is there anything that you could do to help?
- *Collecting feedback*: Has anyone on the team been particularly helpful recently? Gathering this and then delivering it to that person is a simple task but can make people feel appreciated.
- *Sharing a task you're working on*: People are always interested in what their managers are working on, so why not go through it together? Are you writing a job description, discussing the roadmap with your Product Manager, or working on some code of your own? Open the lid on the box and get their opinions. They'll appreciate you sharing and will often have some valuable input.

That's just a sample of some things that you can explore in your one-to-ones—in theory the list is endless. However, the connecting themes are sharing, building trust, and thinking through technical and interpersonal problems. As long as you're spending some time on one or more of those things, you're having a valuable one-to-one.

How to Take Notes and Assign Actions

Throughout the one-to-one, it's useful to jot down notes and assign actions in the shared document. Review the actions at the end of the meeting. Typically I write things directly in the shared document on my laptop. All of our meeting rooms have a TV that you can cast your screen to, so often I'll put the notes up on there as we talk. The notes aren't a secret, after all.

I use a bullet point format to arrange information, and actions are always in bold font. Some hypothetical one-to-one notes with a member of staff called Mary are shown in the following image:

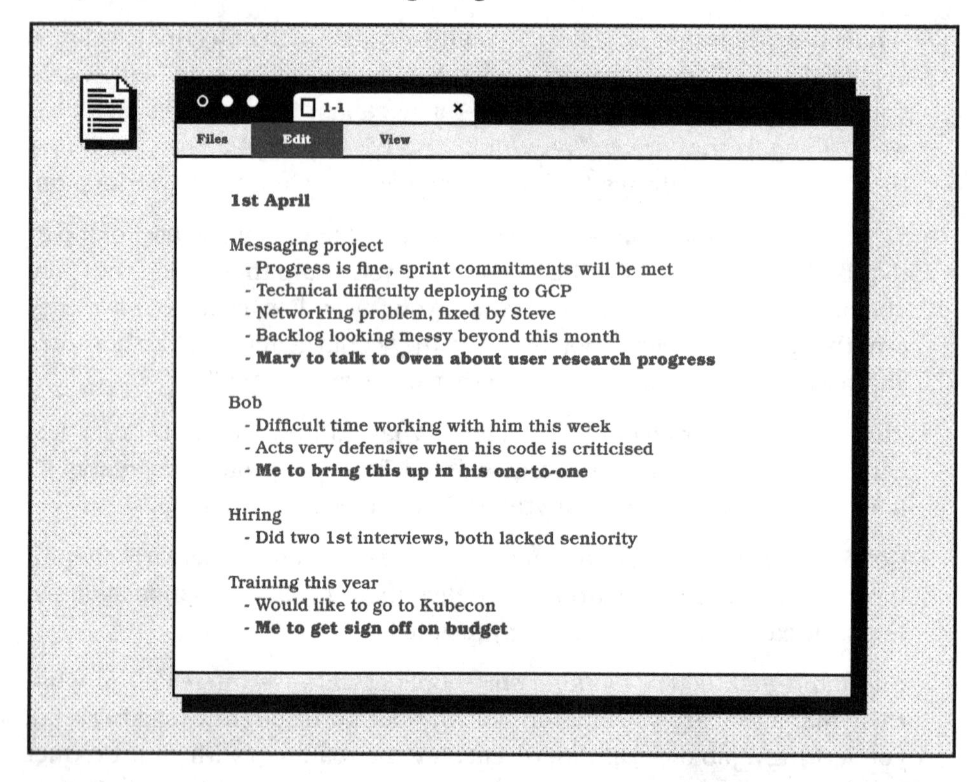

As time passes, and if you make good notes, your one-to-one documents become excellent historical references. You can scan back over them to remind yourself of the topics you were talking about months—and even years—ago.

When you assign actions to yourself, make sure that you put them on your to-do list to action. There's nothing better than having a manager who actually acts on the things that you bring up in your one-to-ones. You'll be surprised how many people I talk to that feel that their manager doesn't.

To help myself remember to do this, my recurring to-do list item for each member of staff's one-to-one has two subtasks:

1. Prepare notes ahead of meeting.
2. Put actions in to-do list.

That way I don't need to think about it. It works for me, so maybe it'll work for you too.

Remember: You Are Not a Therapist

As a diligent, caring manager who is excellent at listening, you may experience your one-to-one sessions turning into therapy sessions for your staff. Although this may be helpful, this isn't something that you're trained in, nor are issues that belong in therapy sessions best placed to be part of your one-to-one meetings. As much as you may want to help, this isn't your area of expertise and you owe it to your staff to find them better support.

If you feel that your staff primarily use your sessions to vent with no solutions proposed, or to talk about in-depth personal issues that aren't related to work, or if you begin to notice signs that you think that your staff are going through difficult times or may be mentally unwell, then there's a limit to which your support can extend.

If you are beginning to worry about your staff, then the best thing that you can do is to raise your concern and ask whether they would be interested in talking to a qualified independent third party. If you're in a company of reasonable size, you may have an HR department that can help. Seek help from your own manager, too, about what to do.

This is, of course, a sensitive area. Despite your intuition, you are (probably) not a medical professional who can make diagnoses. I've been through periods of anxiety and depression and so have my staff, and I have often been able to sense that they may be suffering. However, it's not my place to suggest that. Instead, I have said to them that I sense they may not be feeling their best self and they should make sure that they get some external support if they need it. Once they've opened up about a problem, then it's OK for you both to talk about it, but until that moment, be supportive but refer them elsewhere. Never diagnose.

OK, What's Next?

There you have it: you've got everything that you need to feel confident that you're going to have great one-to-one meetings with your staff.

Let's recap what we've learned in this chapter:

- We've looked at *why* one-to-ones are one of the best opportunities that you have to make an impact as a manager.
- We've learned how to best *prepare* for your one-to-ones, from the time and place that you meet through to how to use a shared agenda.

- For your first one-to-ones with each of your staff, we've explored an exercise called *contracting* that will ensure your relationship gets off to a good start.
- We've looked at a bunch of ideas for the *content* of your meetings, ensuring that you keep well away from boring status updates.
- We've touched upon *how to take notes and assign actions* to ensure that anything that requires a follow up isn't forgotten.

As you spend more one-to-one time with your staff, you'll begin to understand their individual personalities, motivations, and desires. In the next chapter we'll be exploring how your understanding of each individual can be used to ensure that they're doing work that they find suitably challenging, satisfying, and in line with their career growth.

These aren't the droids you're looking for.

> *Obi-Wan Kenobi*

CHAPTER 5

The Right Job for the Person

Last week felt like progress. You sat down with each of your staff and had your first one-to-one meeting with them, using the contracting exercise as a guide. It went down well. They learned a lot about you, and you learned a lot about them. Most importantly, you feel confident that you've got a strong team. Your seniors seem capable and knowledgeable, and your more junior staff are talented and looking for opportunities to contribute and to grow.

Aside from the differences in seniority, you were surprised at the differences in personality. Ben, who was handling your production issue last week, did so with real skill. You were talking to him about some of his most memorable times in industry, and they seemed to be linked by a common thread: chaos and ambiguity. It appears that messy situations fuel his creativity and problem-solving skills.

Amy, on the other hand, is quite the opposite. Despite being similarly senior, and almost single-handedly responsible for building the search infrastructure that powers most of the platform, she can't stand change. Her favorite times at the company were when she was able to focus on difficult—almost academic—distributed systems problems for weeks on end. She described her greatest happiness as when all of her load testing and performance profiling on the search infrastructure upgrade turned out to be completely spot on when it rolled out to production. Her happy place is doing one thing extremely well and being the company expert on it.

You noticed similar disparities among your junior staff. Tara seemed more closely aligned with Ben in that she wanted to learn absolutely everything, from how the UI architecture is put together, through to how the search infrastructure worked, to wanting to get involved in some of the DevOps that support it in production. You could see part of your job would be trying to satiate her desire for newness while ensuring she is making a meaningful contribution to the team. Paulo was more like Amy. He wanted to focus on becoming a deep technical expert on React and found it frustrating when the team were building features that did not allow him much progress toward this goal.

You remember when you were doing the hiring manager training program at your previous company. Like the cubicles and the management style, it was incredibly outdated. Even though you watched it via the HR portal, the static and the scan lines on the video made it look like it was ripped from a VHS. At the end of the embarrassingly scripted and wooden example interview, the narrator had concluded that the techniques that you'd learned would ensure that you would always be able to choose the *right person for the job.*

That particular phrase sat oddly given what you'd learned about your staff last week. Given how challenging and ever-changing the world of technology is, how can you ever pick one specific person that is going to be able to do everything? You'd learned how Ben and Tara thrive in uncertainty, whereas Amy and Paulo crave stability. Each of them wants to grow and work on new and challenging problems. How are you going to ensure that each of your staff are being exposed to the right type of adversity in a way that allows them to enjoy their job as much as they can?

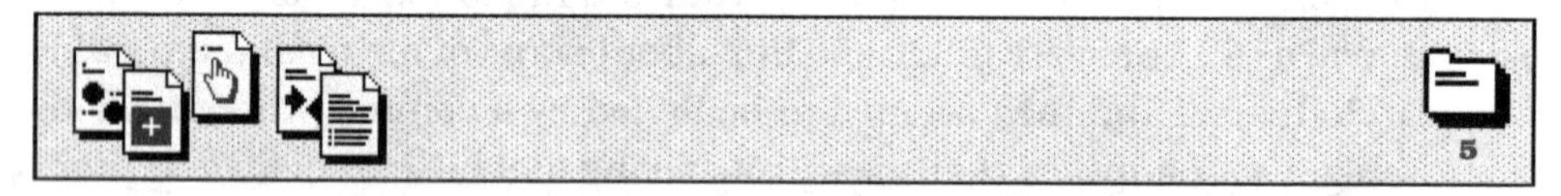

That's what this chapter is going to be about. It's not about getting the right person for the job, it's about getting the *right job for the person.* (When you read the title in passing, did you actually read it as the former?) As a manager, you're going to have staff looking to you to help them achieve their career progression, to ensure that they're working on meaningful problems, and to create or curb the required chaos to do so. You will be the person creating the right jobs for your people to do.

We're going to do a dive into some psychology to see what motivates people, what helps them learn, and how focus and chaos can help and hinder. The concepts that you'll learn in this chapter can be factored into how you frame projects, delegate work, coach individuals, and most importantly, they will ensure that you can understand how best to give your staff the conditions under which they will thrive.

We are going to be taking you on a journey through the following material:

- *What motivates people?* We'll go from basic human needs through to their workplace needs and see how it isn't all about the money (but it helps).
- We'll dip into some *learning theory*, which although typically applied to children, is also relevant to how we seek challenge and learning opportunities at work. We'll see how this reinforces what we previously learned about delegation, what it has in common with video games, and how you can use it on your own team.
- Next, we'll visit *stability and chaos*. We'll look to a classic text on open source software development—*The Cathedral and the Bazaar [Ray01]*—and see how we can reinterpret it to understand why some people love and seek change and why others hate it, and why both viewpoints are right. You'll understand why Ben and Amy are so different.

So, are you ready to join me in getting a bit academic? Fetch your pipe and spectacles and blow the dust off that old psychology book. Let's jump right in.

What Motivates People?

Before we start thinking in depth about strategies to ensure that our staff are challenged, happy, and motivated, let's zoom out and consider the subject at a human level. What is it to be human and what is it to be happy? After all, we've likely heard stories and seen examples of humans all over the world that are at varying—often contradictory—levels of happiness. We've probably seen how often seemingly successful actors and musicians, who society looks at as models of creative and financial success, end up facing bankruptcy or rehab (or both). Likewise, we may have read of monks in Tibet who own no physical possessions other than the robe that they wear, and spend hours a day sitting in silence, reporting that they are as happy and content as they could wish for.

When you think about a time of your life in which you were most happy, where do you place yourself? Were you more or less free than you are now, richer or poorer, a care giver or receiver, student or teacher? What is it about that time that made you so happy?

We live in an economic system and culture that places high value on money and possessions. In fact, we could argue that one of the reasons that people change their job or accept that promotion is that it often brings them more money, which they will see as a catalyst for a better life. Why is it that you were interested in getting into management? Were you motivated by higher salaries or did you want to be a source of guidance to others? You should confront those motivations now, rather than later.

The Hierarchy of Needs

Many factors contribute to humans feeling happy, motivated, and content. One of the most famous psychological models was published by Abraham Maslow in 1943. It's often referred to as *Maslow's hierarchy of needs*. Although it wasn't Maslow himself that structured it as a pyramid (that allegedly came later via a management book), the representation is useful to aid understanding.

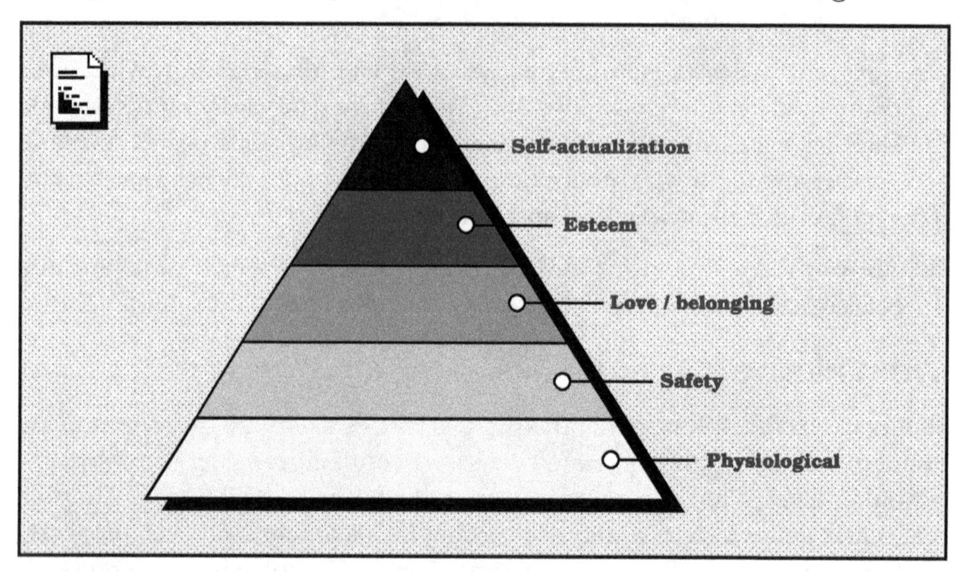

In Maslow's hierarchy, we have five different levels of need, from bottom to top. The largest and most fundamental needs are at the bottom of the pyramid, whereas the higher-level needs are at the top. Each level builds on top of the one below it. Once a level is satisfied, then it enables the level above it to be within reach.

Let's look at each level from the bottom up.

- *Physiological*. The foundational biological needs of a human, such as water, food, warmth, and rest.
- *Safety*. Encompassing all that keeps humans from harm. This includes personal, emotional, and financial security.
- *Love/belonging*. The connection to family, friends, and intimate relationships.
- *Esteem*. Concerning recognition, status, importance, and respect, both of one's self and from others.
- *Self-actualization*. The realization of one's full potential through developing unique talents, skills and abilities, and achieving goals.

The core tenet of Maslow's hierarchy is that humans reach their fullest potential and happiness through the progression to the top of the pyramid. Typically, the bottom two layers are referred to as *basic needs*. The two above that are *psychological needs*. The top of the pyramid is our *self-fulfillment* needs. The argument that can be made is that money and possessions could be reframed as only basic needs, assuming that you have enough to reach the higher levels of the pyramid. An infinite amount of money will not make you happier, but an infinite amount of creative activities and ability to master new skills will.

The Role of Work

Work is a very important part of our lives. After all, you're investing a great deal of your time in reading a book about it. Through our work we are given opportunities to feed into our desires at almost every level of Maslow's model. We can earn money to meet our basic needs, we can work with and serve people to achieve our belonging and esteem needs, and most of all, we can become experienced and skilled, which contributes toward our self-fulfillment.

So, what is it that makes you get up and go to work in the morning? What is it that made you pick up this book? I highly doubt that you wanted to get into management because of the money, because as much as money is nice, there are mornings where I'd rather be warm in bed than chasing the prospect of a 5% pay increase. Additionally, it's likely that you're already working within an industry that is paying above average when compared to other jobs out there. Some more money would be nice, but it probably wouldn't change your life. (I could be wrong though. Perhaps that slightly nicer car will cause you to erupt with joy for the duration of your remaining days on this planet.)

Instead, I reckon the reason that you do what you do, and the reason that you're spending time investing in yourself by reading this book, is because you're a highly motivated individual who is trying to become better in a career that you are passionate about. You want to become even more of an expert. You want to help others. You want to feel the satisfaction of a purpose beyond yourself.

Let's revisit Maslow's model but map it to your life at work:

- *Physiological*. You want to be paid fairly for what you do and to have some reasonable benefits.
- *Safety*. You want job security and a safe working environment.

- *Love/belonging.* You want to work with colleagues that you like and that inspire you, and you want to build good relationships with your team and your manager.
- *Esteem.* You want to be recognized for what you do, both informally through feedback and formally through your job title and promotions.
- *Self-actualization.* You want to be challenged so that you grow and continually expand your skills. You want to feel that there is always somewhere to advance next in your career.

Framed in this way, you can see why those with highly paid but extremely frustrating or boring jobs eventually quit. They are only having their basic needs met. It's also why many will take a large pay cut to start their own business or join a startup. The creativity and autonomy of that work outweighs the lower income.

What's also interesting is that through this model you can begin to see how as a manager you can have a lasting impact on the lives of your staff. Although the basic needs of job security, pay, and benefits are provided by the company as standard, the higher-level needs of belonging, esteem, and self-actualization are something that *you* can have a hand in helping your staff achieve. Isn't that great?

For example, just look at the many ways that you can contribute to fulfilling the higher-level needs of your staff. For *belongingness* needs, you can:

- Build strong relationships with your staff.
- Encourage and demonstrate radically candid feedback to build trust and help them grow.
- Be a role model for those in your team to look up to.
- Connect yourself and your team better to the rest of the organization so that they feel part of something much bigger than themselves.

For *esteem* needs, you can:

- Ensure that your staff are recognized for what they do through appropriate praise and critique.
- Facilitate discussions around career progression, so staff can ensure they have job titles and responsibilities that are accurate and indicate the right level of status.
- Coach and guide staff to improve themselves and become more confident and capable.

Finally, for *self-actualization* needs, you can:

- Ensure that you delegate the right work to the right staff, so that they have continual opportunities to learn, contribute, and grow.
- Provide the environment for staff to learn new skills, both on the job and through attending training and conferences.
- Give staff the freedom to solve problems in the way that they choose, as long as the outcome is correct. This can apply to both abstract problem solving through to concrete decisions around technical approaches and architecture.

These higher-level needs of your staff are where you come in. That's much more than just getting work done: it's contributing toward self-fulfillment. If that's not motivating for you, I don't know what is. Maybe that slightly nicer car?

The Path to Self-Actualization

Every human has fundamental needs. As a manager, you have a great number of opportunities to improve the careers and lives of your staff by ensuring that their needs are met according to Maslow's model. Very few people work in technology just for the money. Often it's because they are innately curious, passionate, and wanting to achieve mastery and self-actualization.

Your Turn: Your Needs

Before we move on, I have some questions for you:

- Have you ever left a high-paying job for a low-paying job? What made you do so?
- Remember the time in your life that you were most satisfied with your job. What was it about the job that made it satisfying and how would you map aspects of that job to Maslow's model?
- What was it about management that made you want to read a book on it? Where does that tie in to your own needs?
- Go through the hierarchy of needs with your staff in your next one-to-one session. How do they feel about their role when considering it? What would help them move higher up the pyramid?

It's your responsibility as a manager to try and create as many opportunities for your staff to meet their needs. Techniques to do so—from delegation, to performance reviews, to career discussions—are all throughout this book. Keep in mind that money and ping-pong tables are not higher-level needs.

Interesting work, challenging technical problems, strong relationships, and candid praise and criticism *do*, however, enable ascension to the top of the pyramid. Get them there.

The Zone of Proximal Development

So, we've had a look at the needs of humans, both inside and outside of work. You can make a lasting difference by helping your staff achieve their higher-level needs. They'll remember that time you were their manager. At the top of Maslow's pyramid is self-actualization: your staff want to define and achieve goals, advance their careers, improve their skills, and become renowned experts in their chosen field. But rather than helping your staff set goals and then leaving it up to chance for them to get there, is it possible to define an incremental path for them to follow? Is there a framework that can guide them on their journey?

We're going to dip into learning theory to see whether we can find out more. Notably, we're going to look at a concept called the *zone of proximal development*, which was a theory originally proposed by Lev Vygotsky in the 1930s. This concept was introduced to understand better how children could be best supported by teachers during their education. During that time, there were contradictory theories and practices on teaching. Schools would predominantly use traditional teacher-led instruction followed by frequent assessments, whereas a fellow psychologist Jean Piaget believed that children learned best when development was self-directed.

Vygotsky saw that whereas this curiosity-led development works well for some subjects—for example, children learn verbal language with little teaching—for most other subjects, the presence of a *more knowledgeable individual* was needed to best advance the child's learning. This was particularly true of mathematics and writing.

The zone of proximal development defines the area in which a person cannot progress without the presence of another person with a higher skill set to assist them. However, once the person has understood and completed the task, they are able to take on tasks of even higher difficulty, pushing their zone of proximal development toward harder and harder tasks. The best teachers and schools, therefore, are able to keep children in their zones of proximal development. The diagram on page 87 illustrates the concept.

In retrospect, this theory almost seems like common sense. However, it highlights two things that are important in order to ensure that learners are continually improving their skills:

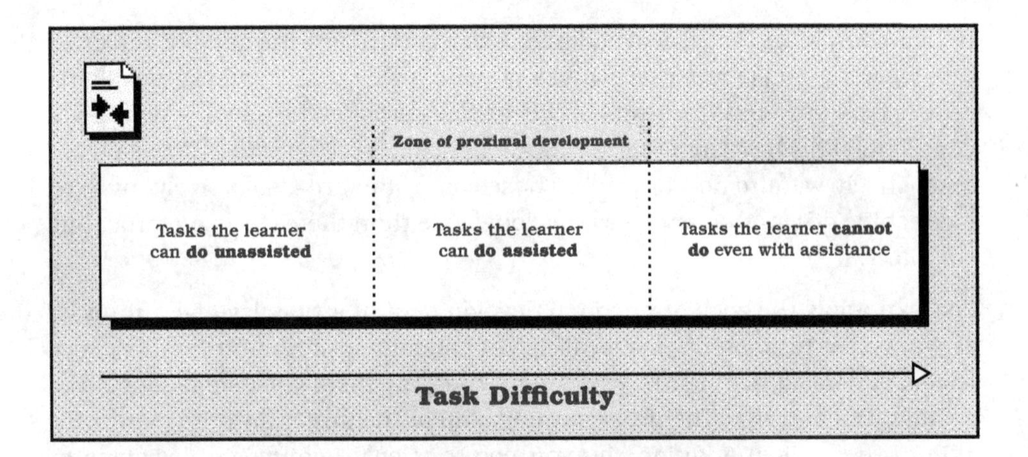

- That they are being *given* the tasks that sit within their zone of proximal development.
- That there are suitable *individuals with more knowledge* to help them complete those tasks with assistance.

Does this sound at all familiar? This doesn't just apply to teachers and children. It applies to all of us wanting to improve our skills at work. But the workplace isn't a classroom. We're not following a curriculum together. Instead, people are working on all manner of tasks at varying difficulties.

As a manager, you can apply the theory of the zone of proximal development at both *task* and *career* levels. For the task level, we can consider it for work within teams, fueling individual growth and encouraging mentorship. At the career level, we can use it as a career-planning methodology for our staff. Let's look at both.

Task-Level Proximal Development

Every task could just be something that needs to get done. Or, alternatively, each task could be viewed as a chance for one of your staff to advance their zone of proximal development. Which is it going to be?

You should give tasks to staff that are in their zone of proximal development. The good news is that you've already read about the tools you need to make this happen. If you squint at the previous diagram, it looks a bit like the delegation diagram that we covered in *Interfacing with Humans*. Delegation can ensure that you give the responsibility for a task to one of your staff while you maintain accountability for it getting done to a high standard and in an acceptably timely manner.

By applying what you've learned about delegation and the zone of proximal development, you can delegate tasks in such a way that encourages collaboration, teaching, and learning, all the while improving the skills of your staff. Excellent teams have staff that are aware of their zone of proximal development and actively want to do work in it. Those same staff are also actively involved in teaching others that are less knowledgeable than them, thus reciprocating the effect.

You can apply the zone of proximal development at a task level in a number of ways. For example, try delegating a challenging programming task to a less-experienced member of your team, but pair them with somebody more senior. This improves the programming skill of the junior and the mentoring skill of the senior. Additionally, you could delegate your own decisions or actions to a member of staff on your team who is interested in becoming a manager in the future. You can mentor them through it, both improving their skills and giving them empowerment.

Your Turn: The Zone of Proximal Development

Before we move on, consider the following:

- Think of each of the staff on your team. What sort of tasks do you think are in the zone of proximal development for each of them? What tasks are beyond their reach right now but could be worth aiming toward in the future?
- Do you have enough support from within the team for your staff to provide mentorship for each other, or are there skill set or seniority gaps? What do you think that you should do when there is a lack of more knowledgeable individuals for particular tasks?
- Is there anyone in your team interested in management? Which of your own tasks could you delegate to them?

Career-Level Proximal Development

Here's where the zone of proximal development gets even more interesting. We can apply the theory beyond individual tasks and instead consider it in the realm of career goals. But first, let's get nerdy.

Have you ever played a role-playing game such as Diablo, World of Warcraft or—I'm showing my age here—EverQuest? A key part of these games is that the player's character gains experience as they are playing the game. This experience allows the character to level up and become more powerful. Additionally, characters often earn skill points that they can invest in themselves to get better at particular actions. You often see these skills arranged in a tree, with the achievement of one skill unlocking the ability to achieve the next, building the character's complete skill set as they progress in the game. There may even be multiple skill trees focusing on alternative progression paths, such as spellcasting, physical strength, and blacksmithing.

The following illustration is a fictitious skill tree for sword-fighting skills for a video game character. In the game, the character would start by learning the initial slash skill, and then they would choose how to invest their skill points from there. They could do so in an exploratory way, but the beauty of the tree representation is that it allows the player to begin with the end in mind. If being an expert at the riposte skill is essential, then they know that they need to learn a basic slash, then invest in parrying, then master the riposte.

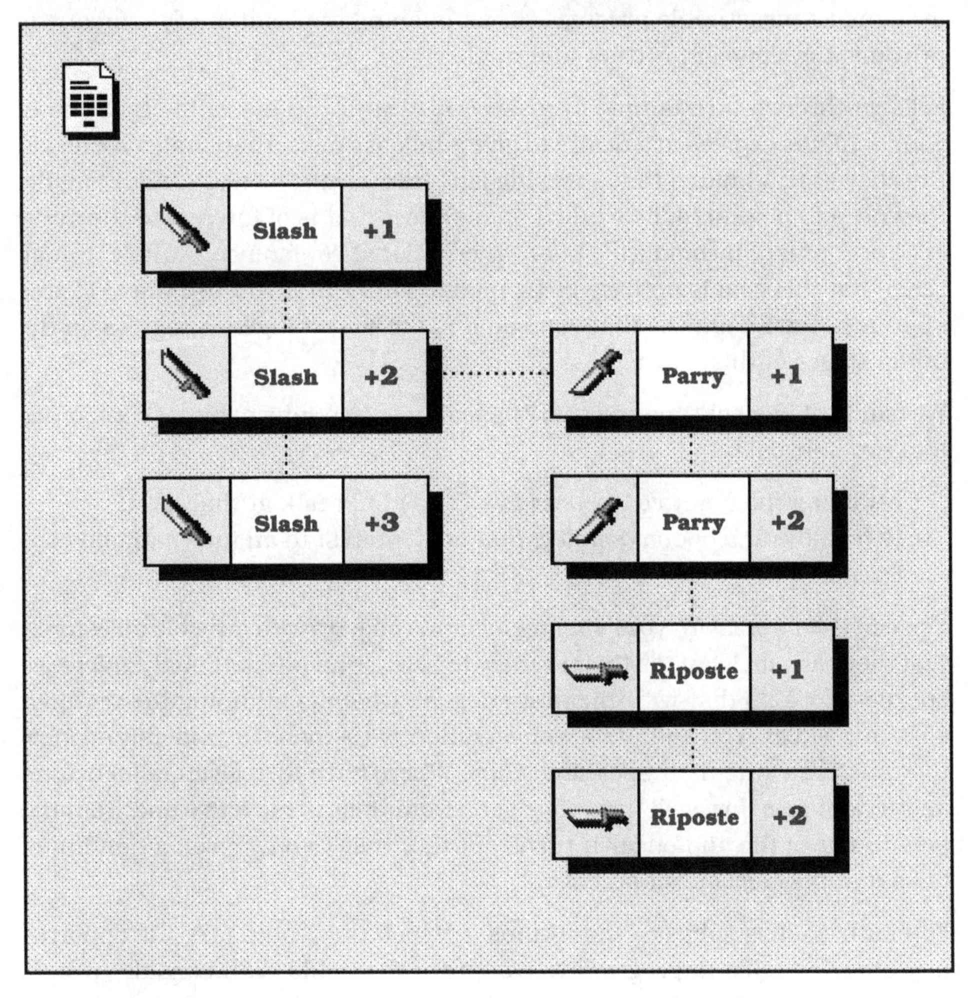

In a sense, we could view this skill tree with the zone of proximal development in mind. A beginner character cannot be an expert at riposting straight away. They must first master the basics and improve from there. Furthermore, what if we weren't thinking about video game characters at all but, instead, were thinking about staff development?

When you spend time with your staff in your one-to-ones, they may talk to you about their desires for their career. Perhaps one day they'd like to be a CTO in their own company, or they'd like to rearchitect the search infrastructure at Google. These are admirable and lofty goals, but in reality, they may be a long way from becoming a reality. However, as their manager, you can work with them to place these career achievements at the bottom of their own skill tree and then plan out the milestones along the way that they can aim for to make measurable progress—thus pushing the frontier of their zone of proximal development further and further.

Let's go through an example to make this clearer. Let's assume that one of your staff has expressed a desire to give a talk at an international conference. But they have admitted that speaking isn't their strongest asset. Additionally, they need to have actually worked on a project that is of interest to the audience and program committee of an international conference. At first, it may seem that this goal is nothing but a pipe dream. But if you use the skill-tree approach, then it may be that you could make some tangible progress on the initial stages of this journey.

We can begin by splitting the goal of speaking at an international conference into two subgoals:

- Attaining the required *speaking skills* to give a talk at that level.
- Being able to *work on a project* that is of interest to an international conference.

You can then work with your staff member to come up with tangible milestones that they can aim for along their journey to become an international conference speaker. Let's do that with the first subgoal, which is attaining the required speaking skills. You initiate a conversation with them to understand how they currently rate their speaking skills. Where have they given talks before, and how often? How well have they gone from their perspective and from the perspective of the audience? It turns out that they've only given a handful of talks and they feel out of practice.

You both discuss subjects and forums in which they could take the opportunity of giving talks to progressively improve their skill. You consider people they could learn from and model. You reason about the order in which they would have to do them so that they can progress in their zone of proximal development. You draw the steps out and come up with a skill tree for them to work toward their first milestone.

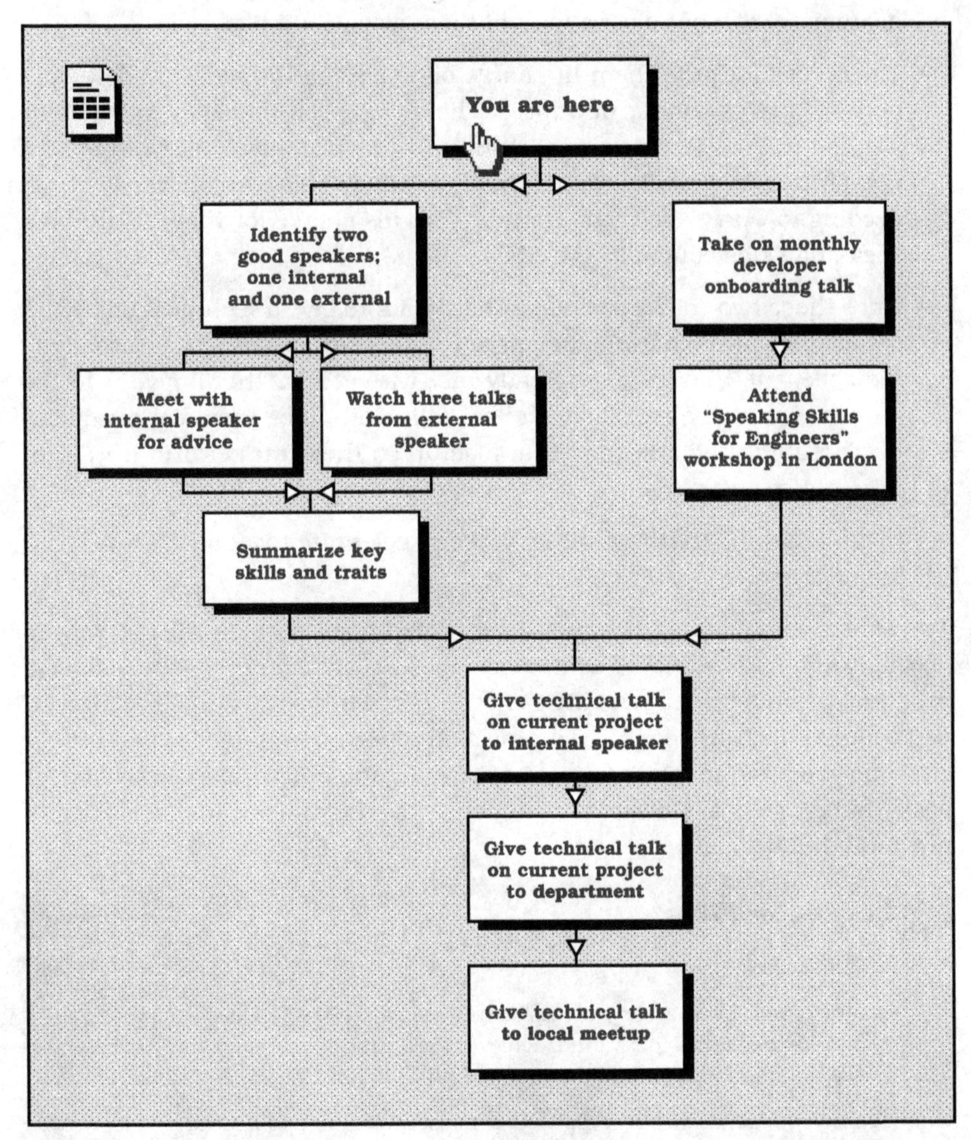

This skill tree gets them from where they are now to the point of being able to speak at a local meetup, which is one step along the way to feeling prepared enough to speak at an international conference. It consists of the following activities for your staff member:

- One fork of the tree involves identifying a person within the company who is an accomplished speaker and one external speaker who has impressed them. For the internal speaker, you agree to introduce them so that they can get advice on giving good talks. For the external speaker, they'll watch some of their talks on YouTube. For both, they'll loop back with you with a summary of what makes both of them give good talks.

- The other fork puts them in charge of delivering the monthly developer onboarding session for new hires. This is material that they know really well, so it gives them an opportunity to practice speaking in front of a new group every month so that they get comfortable with delivery without needing to worry about the content. You also agree for them to attend a speaking skills course in the city.
- Once those two tracks are done, they'll aim to give a technical talk to an increasingly intimidating audience. They'll start by giving it internally to the colleague that they've already met with. Then, they'll give it to the department in the regular technical talk slot. Once they feel confident, they'll submit the talk to a local meetup so they can perform it in front of complete strangers.
- Upon achieving that milestone, you can celebrate together and plan out the next one.

The skill tree shown earlier may have not propelled them all the way to speaking at an international conference. We also didn't address the technical progression required that would be encapsulated in a separate tree. However, you should hopefully see that defining skill trees are both useful and fun for enabling your staff to take a journey of incremental progression toward larger goals, building their skills and confidence by working within their zone of proximal development.

Your Turn: Skill Trees

Some questions:

- Think of your own career progression as a manager. What skills do you wish you had? What goals do you want to achieve? Create a skill tree for your progression toward a first meaningful milestone and run it by your manager for their input.
- Take your staff through the skill tree concept in their one-to-ones and encourage them to think about a goal that they can map out a skill tree for.

The zone of proximal development is a simple yet powerful way to think about development.

- It allows you as a manager to support the growth of your staff by ensuring they work on tasks and projects that allow them to learn new skills with the assistance of others.
- At a task level, you can make choices about how you delegate particular tasks so you can give more junior staff an opportunity to learn and more senior staff an opportunity to mentor.
- At a career level, you can take ambitious goals and then break them down into skill trees that ensure tangible incremental progression.

Ensuring that your staff are continually engaged in conversations about improving their skills and achieving their goals ensures that you are supporting the highest human need in Maslow's model: self-actualization. This is a surefire way of keeping your staff engaged and happy. Who'd have thought that some learning theory and concepts from video games could be so useful?

The Cathedral and the Bazaar

The connective tissue of the internet is built with free and open source software. Data centers run various distributions of Linux. You can browse the web with Mozilla Firefox. It's highly likely your company stores data in one of many open source databases such as Postgres and MySQL. Maybe you have Lucene indexes served up in a distributed cluster by Solr. When you reflect on what we can download and use for free while also receiving continual updates from skilled engineers that contribute for a hobby, it's quite staggering.

As of 2018, the hosting site for software projects using the Git version-control system, GitHub, had received over 1.1 *billion* code contributions from 21 million developers across 96 million projects. Popular open source projects such as TensorFlow, a machine learning framework, have accepted contributions from over 9,000 developers. The React framework for JavaScript has over 10,000 developer contributions. If you were beginning to get into software development today, you may think this is the way open source software has always been done: out in the open with hundreds, if not thousands, of contributors.

However, the watershed moment for this model was the development of the Linux operating system. *The Cathedral and the Bazaar [Ray01]* is an expansion of an essay of the same name that drew attention to how radical it was at the time. Before Linux, many open source software projects were "carefully crafted by individual wizards or small bands of mages working in splendid isolation."

The title of the book comes from the two different software-development models that it covers:

- The *cathedral* model, in which software is free and open source but is developed by the aforementioned hermit wizards. In the book, the GCC compiler is given as an example. These projects have long release intervals and are controlled tightly by the development team.
- The *bazaar* model, in which the entire world contributes and opens their development to the public, despite the chaos. They operate under the tenet that "given enough eyeballs, all bugs are shallow." There's less to lose if occasional issues slip out. Linux is described as the poster child for this movement.

Now, step back from this a second. What has this got to do with management? Well, let's find out. Let's consider the potential motivations of engineers to want to work under both models.

- In the cathedral model, we could reason that the maintainers of the project are most comfortable when they have *control*: that is, they work in a small cabal of level-20 wizards that know each other well, have a final say in their own destiny, and are subject to less distraction and noise from the outside community while they are doing so.
- In the bazaar model, the project maintainers are motivated by *change*: that is, they prefer the fast pace of development by hundreds, if not thousands, of contributors and the fact that all bugs are easy to solve if a large number of people are working on fixing them.

Now, we're not here to debate the relative pros and cons for each model. However, I think that the models highlight relevant individual personality traits and motivations that are well worth thinking about as a manager—that is, that some individuals are highly motivated by *control and stability* and others are motivated by *chaos and change*.

Think of a real cathedral. It is beautifully architected using the finest material and craftsmanship. It is meant to stand the test of time, both in terms of use and the number of years that the building is expected to be on this planet. It evokes quiet contemplation and peaceful reflection. Now think of a bazaar. It is heady with aroma and the voices of sellers. It's bustling and ever-changing. Within the chaos are incredible new things to be found: exotic food and beautiful garments. Those that are driven by new experiences are drawn to the bazaar to be swept away by the experience.

Think of your team for a moment. Given that you've worked with them for a short while, have you begun to see which of them are *cathedral constructors* and which are *bazaar browsers*? How would you categorize yourself? The important consideration is that regardless of whether your staff thrive within chaos or stability, neither is right or wrong. They are just personal preferences and motivations; however, they may change over time depending on a person's life, circumstances, and drive.

The challenge for you as a manager is to have the conversation: identify which of your staff are in either camp, and then build that into how you delegate to them, the opportunities you give them, and the environment within which they work. Let's have a look at both types of people in turn and give some example situations in which they'll thrive. Bear in mind that those who index strongly to either end of the scale may be harder to integrate into teams and may require careful management. That may include moving them to a different team.

Cathedral Constructors

Your cathedral constructors yearn for focus, stability, purism, and craftsmanship. As a manager, you need to be able to offer them the following opportunities:

- To become *subject-matter experts*. Is there the opportunity for them to build and contribute to a core area of the technology stack? Can they be given ownership over it in such a way that lets them become the most knowledgeable in the company about that particular part?

- To be a *cornerstone* of the team. Is there an opportunity for them to become the individual that their teammates seek advice from when approaching something new? Are they able to provide the scaffolding that new functionality is built on top of?

- To go *deep, not wide*. Your cathedral constructors relish in thorough knowledge. Is there an opportunity for them to contribute back to open source projects that they are an expert in? Are they able to purposefully stay away from newness in order to practice deep mastery?

- To *show others their own ways*. Those that become experts also benefit (and often relish) in teaching others what they know. Ensure that your cathedral constructors have the opportunities for mentoring.

- To *revel in the detail*. Try to keep incomplete knowledge or politics away from your cathedral constructors so they can focus on the small details. Let them be in their happy place.

Bazaar Browsers

The bazaar browser wishes to sample the sights and sounds of as much as possible. They want excitement, variety, chaos, and change. This is where they thrive and grow. You need to be able to offer them the following opportunities:

- To get *as much newness as possible*. Is there a new project coming up? Your bazaar browsers want to sample new challenges, domains, products, and technologies as much as possible. Put whatever new thing you have in front of them first.
- To *build and throw away*. Are you about to go through a phase of quick prototyping and iteration? Are you going to be experimenting with some new technology? Your bazaar browsers love this. Send it their way.
- To *avoid stagnation*. Unlike the cathedral constructors, the bazaar browsers find working on the same area for a long time a sign of stagnation rather than mastery. Ensure that you're talking with them about how they feel, and give them opportunities to switch up what they're doing as time goes on.
- To *even move around teams*. If your team is unable to keep a bazaar browser motivated, then why not consider loaning them out to another team for a short while? They'll get to do something new and fun and bring back knowledge. You could even take on a bazaar browser in return to inject some new energy into your team.

Having the Conversation

While many may have had a discussion with their manager about whether they want to become managers or not in the future, one could speculate that far fewer have discussed whether they are cathedral builders or bazaar browsers. This is great material for your one-to-ones.

You may also find that your preconceptions about some of your staff could be completely wrong: maybe Bob was underperforming recently because he needs more variety, but you thought it was best for him to stick with what he knows. That was wrong! Maybe Alice is happiest when she can stay inside her comfort zone, and the new challenge you gave her is uncomfortable rather than exciting.

By knowing what drives people, you can better place them in upcoming projects and teams; both for their happiness and for the best performance for your business.

Your Turn: The Cathedral and the Bazaar

So, before we wrap up this chapter, some questions for you:

- Where do you see yourself with regards to cathedral constructors and bazaar browsers? Are you motivated by change, or do you crave stability? Which projects have been your favorite, and is it because of your preference there?
- Bring this subject up with your staff in their one-to-ones. Which of your staff identify with either trait? Are you surprised by what their answers are based on your preconception of them? What work is coming up that you can best involve them in?

A Review Before Reviews

There we have it: you've learned that you have a duty as a manager to ensure that you're giving your staff the best possible opportunity to be happy, motivated, and able to grow. It's not about finding the right person for the job. It's about finding the *right job for the person.*

Most of the time, this task is achievable with time. You can help people find their groove. However, there are times when this is an impossible task given a particular member of staff and a particular team. We'll explore more about what happens when individuals don't work out in *Game Over.*

Here's what we covered:

- We learned about *Maslow's hierarchy of needs* as a model for exploring what motivates human beings. Money and possessions occupy the lower part of the pyramid, whereas self-actualization goals—the ones concerned with the fulfillment of human potential—are at the top. When mapping this on to the workplace, you can see that you have many opportunities to contribute to fulfilling the higher-level needs of your reports.
- We then explored the *zone of proximal development* as a way of understanding how people improve their skills. We saw how it forms the core of effective delegation and mentorship. We also saw how you can push the proximal development frontier via skill trees, thus enabling your staff to continually build toward their career goals.
- We finished by learning about *cathedral constructors* and *bazaar browsers,* two different but equally important personal traits that your staff may have. We saw how you can satiate those that fall into either camp.

You've now got a lot of material to talk about in your one-to-ones. Make sure you do so. Next up we're going to be diving deep into a special one-to-one that usually happens twice a year: the *performance reviews*. But don't worry, with some guidance you'll learn to look forward to them.

Life can only be understood backwards; but it must be lived forwards.

Søren Kierkegaard

CHAPTER 6

The Most Wonderful Time of the Year

What's that familiar feeling of worry as you check your calendar for the next few weeks? Ah yes, that's right: it's review time. You remember your first performance review at your previous company. Those butterflies in your stomach as you stood outside of your manager's office. The look of dread on the face of your colleagues as they exited the room. Your sweaty palm as it touched the cold door handle. It was awful. Entering the room, you sat down in the chair in silence while you waited for your boss to speak.

"So," he said in an inquiring tone. "How do you think you've done this year?"

"Erm, pretty well, I guess."

"Interesting," he replied. A long pause followed. It hung in the air.

It was probably one of the most uncomfortable hours of your working life. It felt like an interrogation. You had your view of the last six months extracted out of you by an individual playing both good cop and bad cop simultaneously. You ran through the main features that you worked on. You thought they all went pretty well, but all he did was find holes to pick in them. You talked through some of the areas in which you could have done better. There was no comment. At the end of the meeting you had no idea whether you'd done well or not.

"Oh, and one more thing: 3% increase this year. It'll be in your January pay."

You catch yourself daydreaming and bring yourself back into the present moment. You look at your calendar and realize that it's not just one review you need to sit through this time around. It's *seven*. And *you're* the manager. Oh dear. Do you really have to adopt that horrible persona and drag your staff (and yourself) through hour after hellish hour? Surely there has to be a better way...

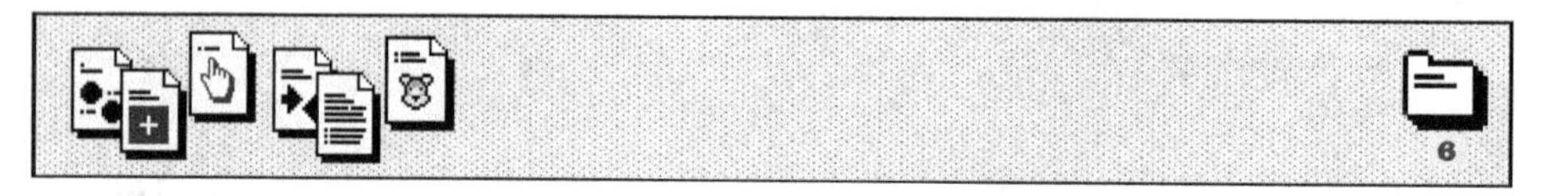

Well, there is. Performance reviews need not be painful. When done well, reviews can be reflective, full of praise and constructive criticism, and act as a springboard for the future. They can further strengthen the relationship between you and your staff. As a manager, you're on the other side of the table, so you have the opportunity to deliver valuable sessions for your staff. And it just so happens that you're in luck, because that's exactly what this chapter is going to be about. We're going to give you everything that you need to know about giving great performance reviews, from preparation to the meeting to what happens afterward.

It can be challenging to be in the manager's seat for the first time. But you should see this as an opportunity to be the manager that doesn't fill their staff with dread when it's review time. The good news is with some knowledge and careful preparation, you can become a master at doing performance reviews. With time, you may even look forward to doing them, since they're an opportunity for a deep conversation about personal development and goals.

But let's get one thing straight: very few people like performance reviews. They're essential yet often unpleasant—a trip to the proverbial dentist. Yet, despite the unpleasantness, they're the best opportunity that you have to push your top performers further and course-correct those that are underperforming. Use them well, and your staff will only get better. Use them badly, and you'll be in for some very awkward conversations.

Here's what we're going to cover in this chapter:

- *Myth busting.* First off, we're going to break some common misconceptions and myths about performance reviews.
- *Preparation.* We're going to break down everything that you need to do before the meetings into easily implementable chunks. Since you'll be managing lots of documents and emails in the lead up, we'll give you a simple tracker system to keep on top of everything that's in flight.
- *The written review.* We'll explore how the written review should be structured, including how much you and your staff should write. You can use the template in this chapter for your own reviews if you wish.

- *Peer feedback*. Getting a broad selection of opinions from others is just as valuable (and sometimes *more* valuable) than the feedback that you give your staff yourself. We'll look at how to collect it and you'll get an email template to use.
- *What to do on the day*. Then we'll cover the meeting itself, making sure that both you and your staff get what you need from the conversation.
- *Money*. This part of the review is often handled badly. We'll show you how to cover this tricky part of the conversation so that it lands well for both you and your staff.

So, are you ready? You'll realize that reviews aren't that bad after all. Let's get started.

Myth Busting

Before we go any further, we're going to consider some misconceptions about performance reviews. Part of the reason that reviews can suck is because both parties are going in with a mindset that hinders rather than helps. Let's have a look at some incorrect assumptions that people can have around performance reviews—from both the reviewer and the reviewee—so that you can understand how to not repeat the mistakes of others.

Myth 1: Reviews Are for Managers to Give Top-Down Feedback

A review is not a one-way debrief. It's not the parole board at the prison giving their rubber-stamped verdict on whether an inmate can walk free. Reviews should be a *two-way* process. In the weeks leading up to reviews, both you and your staff will put in the time and effort to prepare. They will reflect on their performance and so will you. You will collect peer feedback for them so you can discuss it in partnership with them. They will be given the space and opportunity to talk about their goals for the future.

Remember that in the review process you, as a manager, are the *facilitator*. You are not the dictator. Deliver a stellar service to your staff, not a judgment.

Myth 2: Reviews Are Just a Thing That the Company Does

Yes, it may be the case that it's your HR department that sends out the notifications when it's time to prepare for performance reviews, but it does *not* mean that reviews are just a thing that you should unwillingly do to check the box. Performance reviews are an essential part of your toolkit as a manager to ensure that your staff are supported, given opportunities to talk in depth about their careers, and to continually set goals. Even if the company was not making you do your reviews, you should do them anyway. In fact, if they only mandate them happening once a year, then you should probably do them every six months regardless. We'll touch on the frequency of reviews shortly.

Reviews are absolutely worth your time. You should engage fully with the process and also make sure that your staff do too. This is because:

- They are the best opportunity to set the bar for what you expect for the next six months.
- They are your best chance to apply course correction if your staff aren't going in the right direction.
- They build trust and rapport through their introspective nature.

Myth 3: Reviews Only Really Matter for Underperforming Staff

That's not true. In fact, it's quite the opposite! Reviews are not there to make sure that staff that aren't performing get the proverbial kick that they need to do better. Reviews are for *everyone*. Each of your staff deserves the chance to sit down with you and contemplate where they've been, where they are, and where they're going. The people that benefit from reviews the most are your highest-performing staff. By giving them the time and space to explore their career, combined with your support and input, you can ensure that your superstars continue on their high-growth trajectory.

Reviews are not for nitpicking negatives. They are for praise, critique, planning career goals, and dreaming about the future, and your best performers need your time and energy more than anyone else.

Myth 4: Since People Hate Reviews, Get Them Over and Done with Quickly

Wrong! People hate reviews because they are often done badly, not because reviews themselves are bad. Don't assume that everyone hates reviews. Also don't assume that even if people hate reviews that they'd rather skip them or do a cut-down version of them in order to check the box and move on with their lives. Engage fully. Always give reviews your utmost attention and prepare and deliver them with the same detail and passion that you give to the projects that your team is working on.

Part of your challenge as a manager is to turn even the most skeptical members of staff into ones that actively look forward to reviews. This chapter will show you how.

Myth 5: The Review Should Be a Surprise on the Day

Five minutes before the review, your member of staff should not be sitting there nervously, staring at the meeting room, filled with dread about what is going to be revealed. In fact, it should be the opposite. Most of the time that goes into performance reviews happens before the actual meeting. You'll see how both of you will contribute to a written document that will allow you to record your thoughts in a concise and considered way and allow plenty of time for your opinions to be exchanged beforehand. You'll also see how to collect and compile peer feedback so that your member of staff has a more complete picture of their performance to discuss.

When it comes to the meeting, the pertinent information should have already been expressed, chewed over, and understood. It should never be a surprise.

Myth 6: Reviews Are Used to Deliver Pay Raises

Reviews should not be the grand unveiling of pay increases. It's time for you to decouple the delivery of pay and performance, despite the fact that the two are conceptually coupled. Keep the performance review about the performance only. You'll see why this is beneficial to both you and the person that you are reviewing, but you'll also see how it opens the door to there being other more sensitive ways of delivering salary adjustments.

Myth 7: Reviews Are the Only Place Where Performance Is Discussed

A common bad practice is to use performance reviews as the only place where discussion about performance happens. What you actually want is for you and your staff to have open and transparent conversations about performance—good and bad—all throughout the year. The emphasis of performance reviews should be on the *review*. They're a checkpoint to talk about the time period that has passed and the one that's coming up next. Yes, that does involve talking about performance, but it should only be one of the *many* times that you visit the topic as the weeks and months go by. To an extent, there should be no surprises in what you discuss with your staff, and vice versa.

How to Prepare for Performance Reviews

Now that we've addressed a selection of common misconceptions about performance reviews, let's get to business. In this section you're going to learn exactly how to prepare everything that you need to make your performance reviews a success.

The first question to address is this: how often should performance reviews be happening? I recommend every six months, once at the end of the year (say, around December) and once in the middle of the year (say, around June). A few reasons support this:

- Waiting an entire year between performance reviews is way too long. Conversely, a review every quarter feels too often. Six months is a long enough period of time for you and your staff to have done a substantial amount of work to reflect on and for everyone to have progress toward previous goals.

- Two checkpoints a year increase the chances you have to course correct anyone that needs assistance with their performance or direction before it's too late.

- The end-of-year review will most likely have the pay element looming, although we'll look at some techniques for dealing with that. The mid-year review allows a touchpoint purely around performance without having to worry about the added complexity of pay.

Even if you find that your company only does performance reviews at the end of the year, there should be absolutely nothing stopping you from doing them with your team during the middle of the year as well. After all, you're only trying to do a good job of running your team!

So, to the next question: how much preparation do reviews need? The answer is quite a lot. However, the more you do them, the more efficient you'll get. Additionally, you're about to learn a simple step-by-step process using a tracker to make it easier to organize yourself.

I would advise getting started on the preparation around three weeks before the reviews. This is because:

- Often, writing reviews can take a fair bit of time, especially if you or your staff are not natural writers or need to take a long hard think about what they want to cover.

- You'll be asking for peer feedback for each of your staff, which you'll need to request, collect, and then present. Regardless of how smart and talented your colleagues are, the peer review process always takes longer than you'll expect. Allow time for it to happen so that each of your staff has a good selection of peer review to digest.

- No doubt there will be production issues, bugs, things that catch fire, and vacations to contend with. Starting early builds in contingency for all of this.

Review Tracker

	A	B	C	D	E	F	G
1	Staff	Joyce	Ianthe	Finch	Peter	Alan	Rebecca
2	Form shared						
3	Mine filled in						
4	Theirs filled in						
5	Peer Feedback 1						
6	Peer Feedback 2						
7	Peer Feedback 3						
8	Peer Feedback 4						
9	Peer Feedback 5						
10	Meeting booked						
11	Feedback collated						
12	Goals						

Now, to the tracker I mentioned. Let's create your own, see the sidebar on page 106.

You can tackle one task straightaway and get it ticked off: booking each of the meetings. One strategy is to use your existing one-to-one slots you have booked with your staff, as reviews can fit well into an hour if the preparation has all been done up front. You can go into your calendar software and for the week you're going to be doing the reviews, just edit the name of the meeting. That way you don't need to play calendar Jenga to find even more free space.

Create your tracker, book your meetings while ensuring that they're in a private room, and let's move on.

Getting Peer Feedback

If you enjoy herding cats, then you'll enjoy gathering peer feedback. Your company may have an internal HR system that solicits peer feedback, but there's nothing wrong with using good old email for the job.

Your Turn: Create Your Review Tracker

Create the following layout in the spreadsheet software of your choice. Here's what each of the columns mean:

- Staff: The names of each of your staff that you'll be reviewing. Simple.
- Form shared: Each cell is marked "yes" when you've created and shared the blank review template form that we'll be looking at in the next section with each member of your staff.
- Mine filled in: Each cell is marked "yes" when you've written all of the relevant sections in the review form that you have to for that member of staff.
- Theirs filled in: This cell is marked "yes" when the member of staff has completed writing their sections of the same form. Since it'll be a shared document, you'll see when they're done.
- Peer feedback 1–5: These cells should begin colored red since you haven't sent the peer feedback yet. However, you'll be identifying up to five members of staff for each of your reports that will be sent peer feedback requests for them. You'll change the color to green when you've received it back.
- Meeting booked: When you've got the performance review booked into the calendar, you'll mark this as "yes."
- Peer feedback collated: When all of the peer feedback has come in for each member of staff, you'll be copying it all into a single document for them to review, and marking this "yes."
- Goals: After the review has finished, you'll firm up and mutually agree the goals that they will be working on for the next six-month period. And guess what—after you're done, you'll mark it "yes."

In your tracker grid, for each member of your staff, select three to five people to give feedback to. Aim for the feedback to come from as many different viewpoints as possible. For example, you can pick from:

- Their colleagues within the team.
- Anyone external to the team that they work with regularly (such as someone who they often seek advice from or who helps review their code).
- Anyone who they regularly spend time mentoring on another team.
- Colleagues with different skill sets that they interact with often who can bring a different angle to their performance, such as product managers, product marketers, your own boss, and so on.

Don't just pick a couple of their closest teammates. Consciously think about how to compose a picture from multiple angles that will bring them real value when they read it.

Your Turn: Peer Feedback Emails

Once you've selected the teammates that are going to give each of your staff peer feedback, put them in your tracker sheet. Next, for each person, you're going to send an email asking for feedback. Feel free to use this template.

Hello,

If you're receiving this email then it's because I would love to get some peer feedback on the person mentioned in the subject. You've been picked because you either work very closely with this person or you are managed by them. Ideally I'd like this feedback in the next two weeks.

Peer feedback is an extremely insightful tool for people in their reviews. Typically, the content of a review between a member of staff and their line manager shouldn't be that much of a surprise. However, the really interesting data comes from others in the organization, as the reviewee can form a fuller picture of how they're performing in their role.

I'd like you to try your best to answer the following questions:

- What has this person excelled at in the last six months?
- What should this person do differently and why?
- What other feedback would you like to give?
- Would you like to give this feedback anonymously?

Just replying to this email is fine with me.

Change the background of each of the red cells to yellow to indicate that you've requested their input.

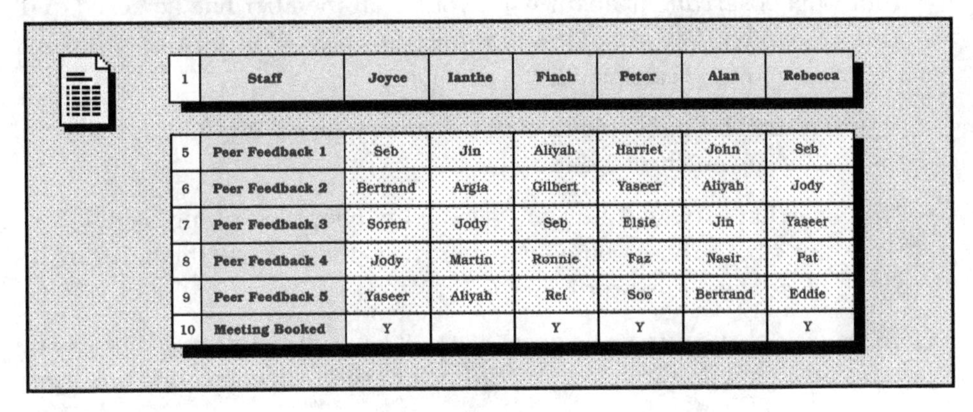

1	Staff	Joyce	Ianthe	Finch	Peter	Alan	Rebecca
5	Peer Feedback 1	Seb	Jin	Aliyah	Harriet	John	Seb
6	Peer Feedback 2	Bertrand	Argia	Gilbert	Yaseer	Aliyah	Jody
7	Peer Feedback 3	Soren	Jody	Seb	Elsie	Jin	Yaseer
8	Peer Feedback 4	Jody	Martin	Ronnie	Faz	Nasir	Pat
9	Peer Feedback 5	Yaseer	Aliyah	Rei	Soo	Bertrand	Eddie
10	Meeting Booked	Y		Y	Y		Y

Then, you wait. And sometimes you wait and wait. As each piece of feedback comes in, mark the corresponding cell green. Some people are notoriously bad at replying to non-urgent emails, and you may find yourself waiting forever. I usually wait a week and then send a polite nudge out to everyone that hasn't replied yet. Don't expect that you'll always get a reply. That's OK. People are busy and it's the reason that we went broad with five people per person in the first place. Individuals may be on vacation, or snowed under with a particular project, so do be respectful that you're asking them for a favor. One nudge is polite, but any more is just annoying. Turn to others if any of your staff is lacking a broad range of opinions.

You may find that you receive peer feedback that is nothing more than a few one-liners, seemingly written in a couple of minutes. If this happens, try to reply and ask if they can expand on their thoughts, either in a face-to-face conversation or via another email. Sometimes the peer review process is seen as another box-ticking exercise, but you should try to make sure that each person gives meaningful and actionable insights.

The Written Review Form

The written part of the review is the sun that the review orbits around. It contains all of the meaty content. Your company may already give you a review form to use. If so, then that's absolutely fine. Go ahead and use it. However, if your company doesn't give you a form to use, or perhaps you're not satisfied with the one that you're being given, then you can use the form provided in this chapter that I've been using for a number of years. It has served me pretty well, and it may serve you well also. You should prepare the review in a shared document so that both of you can see what the other has written.

The review form is broadly split into three parts:

- Your *own observations* about what your staff member has achieved and areas in which they should seek to develop next. You fill this in before the review so they can read it.
- Your *direct report's observations* on the same topics. They fill this in before the review so you can read it.
- The *goals* that you have mutually agreed upon for the coming six months. You think about these going into the review and discuss them during it. You finalize them afterward.

Before we look at the questions specifically, it's worth reiterating that two of the three elements are written and read before the review meeting has even happened. That's extremely important. Not only is it helpful to you both to have taken the time to digest everything that has been written, it also helps ensure that you both enter the meeting with the right mindset in order to have a constructive conversation.

As the person on the receiving end of the review, it's deeply unpleasant to turn up with no idea of the direction that the meeting is going to take, especially if it hasn't been a stellar year for them. These meetings are not an occasion for a big reveal, and that's true for both good and bad news. It also goes without saying that the bad reviews are much harder to stomach than the good ones. Delivery of critique, especially when there's a lot to criticize, can put people in a spin. By sharing what you've written for them beforehand, you give your staff time to mentally prepare.

To frame why this is useful, consider the five stages that people tend to go through when receiving bad news:

- Ignore it
- Deny it
- Blame others
- Assume responsibility for it
- Find a solution

Staff reading the document before the meeting have the chance to move through steps 1–4 in their own minds. This makes the meeting focus on step 5, which is a much more productive use of both of your time.

Right, let's start putting together your review form. We will step through each of the questions in turn. Remember that writing reviews takes time. Allow your staff time to get them done. One way to find this time is to treat it like any other work in your team by putting tickets for each of them in the backlog.

Their Questions to Answer

The initial part of the review is for your staff to fill in. It prompts them to consider:

- Their *achievements* over the period of the time since the last review.
- A *reflection* over the time period for their role, projects, and career.
- Areas that they would like to *develop* in the coming period.
- Thoughts about the *future* of their role from where they currently stand.

Let's look at each of those questions in more detail.

Achievements

Question: *Think about your achievements, big or small, over the last six months. What have you accomplished that makes you proud and why?*

This kicks off proceedings on a positive note. Do bear in mind that many people find it difficult to talk about their own achievements, or may downplay them, so this is why there is a similar question for you to answer later. However, it's always insightful to see what a member of staff thinks about their own performance.

For example, you're able to see whether they hone in on their own achievements or those of the team. It gives you an idea of what *they* see as their highlight reel compared to what *you* see. Additionally, it gives you plenty of material for your conversation in the review where you can dish out praise.

Reflection

Question: *How do you feel in your role right now? Has that changed for the better or worse since your last review?*

After the achievements comes introspection. It encourages them to step back and consider their place in their career and the company and ponder where that aligns with their expectations. You shouldn't expect all of your staff to craft philosophical essays here.

Recalling our discussion of cathedral constructors and bazaar builders in *The Right Job for the Person*, it may be the case that a member of staff is absolutely fine where they are now and find it difficult to comprehend why anyone would ever feel otherwise. That's absolutely fine. Conversely, this question may also open an existential can of infinite worms. That's also absolutely fine. It's great material to discuss in your meeting.

Development

Question: *Is there anything that could have gone better during the period? What skills do you think that you need to develop?*

After the internal reflection, they're prompted to think about what they'd like to develop. These may range from technical to interpersonal skills. Again, it also allows you to better understand their view of themselves, which you can contrast with your own and that which was written about in the peer feedback.

The Future

Question: *Consider your future at the company. Where are you aiming to go? What would you ideally like to accomplish or work on? Or are you happy where you are now?*

This is a hard question for even the most goal-focused person. Do remember again that your cathedral builders may be perfectly happy where they are. Your bazaar browsers may want to be anywhere but here, and that isn't necessarily something that you should feel betrayed about: they are just trying to satiate their own curiosity and hunger for newness. But remember that if your staff are happy where they are now, then that's something to be celebrated. They've found their fit.

Support

Question: *What sort of support do you need in order to get to where you want to go in future?*

Your staff may just need time to gain the expertise that they want. In the worst case, they might not even be able to get what they want at your current company: perhaps consider what you would do if their real passion was C++, but you're a Java shop. Maybe you'll need to think about requesting some budget to send them on a specialist training course. Perhaps their trajectory might eventually end up with them joining a different team. Either way, you should consider their needs carefully to consider how you can plan to get them where they want to go.

Your Questions to Answer

The second half of the review is for you to write. For each of your staff, you should consider:

- Your own view of their *achievements* over the period.
- Where you think that they should focus on *developing*.
- The themes that you noticed in their *peer feedback*.

Again, let's see these in more detail.

Achievements

Question: *What has this person done in the last six months that has impressed you? What do you consider their biggest achievements?*

You get to give your own opinion on their accomplishments. Despite you writing this in a shared document, you should resist just repeating what they've written for this section. Instead, try and offer your own unique view on their performance. What have you observed that maybe they haven't? Have you noticed that they have a lot of excellent traits that come naturally to them that they don't otherwise notice? Say so. You're also able to view their performance through the lens of the whole team, so build that angle in also.

Development

Question: *Where do you think that this person should focus on developing? What traits and skills could they work on and how can you help them get there?*

This section isn't called "failures" for a reason. Even staff that are struggling need to have critique framed as development. Try your best to think of potential solutions for how they can improve and how, as their manager, you can help them achieve those improvements. You may find that this section is hardest when your staff are doing exceptionally well, since you'll have to think hard about how to push them further. It's especially important to do so since your stars thrive when challenged.

Peer Feedback

Question: *What were the key themes from this person's peer feedback? Does it reinforce what you both already know, or are there any new observations?*

Review the peer feedback that you've collected for them. Is it good, bad, or mixed? Are there recurring themes that are worth discussing? Is there a piece of feedback where it would be worth finding out more from the individual that wrote it?

Gather all of the feedback together verbatim in another shared document and link to it in this section. Occasionally there may be feedback that you'd like to edit in some way—perhaps because it is overly critical or could be interpreted badly—so do so. However, in practice I've only experienced a few occasions over the years where I was concerned about presenting it as it was sent.

Getting Them Written

Don't underestimate the amount of time that it takes to write reviews. In the weeks before reviews, you'll need to organize your time so that you have plenty of uninterrupted focus to get your words down on paper. From my own experience, it usually takes me an hour per member of staff, but reviews with more pertinent content—both good and bad—can take longer.

I'm a big believer in transparency, so I always write the reviews directly into the shared form where they can see them. That way we can leave each other comments as we write. Once you've written your part, let them know that you're done and they can read it if they haven't seen it already.

Try to get them written with plenty of time before the review. This ensures that you've both had time to read, digest, and prepare what you'd like to talk about, which ensures that the hour that you spend together is as valuable as you can make it.

Once you've written all of your parts and your staff have written theirs, your tracker form should look a bit like this before you have the actual meeting.

Review Tracker

	A	B	C	D	E	F	G
1	**Staff**	**Joyce**	**Ianthe**	**Finch**	**Peter**	**Alan**	**Rebecca**
2	**Form shared**	Y	Y	Y	Y	Y	Y
3	**Mine filled in**	Y	Y	Y	Y	Y	Y
4	**Theirs filled in**	Y	Y	Y	Y	Y	Y
5	**Peer Feedback 1**	Seb	Jin	Aliyah	Harriet	John	Seb
6	**Peer Feedback 2**	Bertrand	Argia	Gilbert	Yaseer	Aliyah	Jody
7	**Peer Feedback 3**	Soren	Jody	Seb	Elsie	Jin	Yaseer
8	**Peer Feedback 4**	Jody	Martin	Ronnie	Faz	Nasir	Pat
9	**Peer Feedback 5**	Yaseer	Aliyah	Rei	Soo	Bertrand	Eddie
10	**Meeting booked**	Y	Y	Y	Y	Y	Y
11	**Feedback collated**	Y	Y	Y	Y	Y	Y
12	**Goals**						

What to Do on the Day

The good news is that if you've done all of your preparation, your performance on the day should be relatively straightforward. Before you go into the room, reread the review form and peer feedback and think about the items that you think you definitely should discuss. Using the shared review form, you've already communicated a great deal: both of you have written your thoughts and feelings and digested them. Remember that even if your feedback was critical, they've had that opportunity to assume responsibility for it. Don't dwell on the negative side of the critique. This meeting is for optimistic discussion on how they can move on through it.

A big mistake is to spend the meeting just stepping through everything that you've listed in the written review. Don't do this: it's not a good use of either of your time. Instead, try and split the meeting into three parts of equal size:

- A reflective discussion about the period that has just passed.
- A forward-thinking discussion about the future.
- A collaborative session drafting goals that they can achieve.

In the first two parts, keep the discussion loose and let them do the talking. Ask questions and nudge them along to explore how they feel. Try not to presuppose or suggest too much. This meeting is not where they need to have a revelation about themselves, but instead it's a milestone in a continuous process. You should both be continuing to have these discussions about their career in your one-to-ones.

For example, here are some leading questions that you could ask in reference to the content of the written review form:

- What is it about that achievement that made you feel so proud? Do you feel like that often, or was this a unique occasion?
- You've mentioned that you want to develop that particular area. How do you think you can do that, and how can I support you in doing so?
- How do you feel about your team? Who in particular do you most like working with? Is there anyone that you find harder to work with? Why is that?
- How challenging is the work that you're doing now compared to the last few years? Are you getting enough opportunity to stretch yourself and to learn new things?
- How does your current role compare to where you predicted that you'd be five years ago? Why is that?
- What do you think of *my* performance as your manager? What can I do to support you and the team better?

Regardless of the actual questions, use some simple guiding principles:

- At this stage, prompting their reflections is more important than you giving yours. You've done your part in writing. This is where they begin to figure out what it all means and where they would like to go.
- You're not there to figure out their future career for them, but you're there to listen and facilitate the conversation.
- Keep the thought bubble over their head.

During the last third of the review, begin drafting some goals for the next six months together. You don't need to complete them in the review, but you should begin the thought process collaboratively. What would they like to achieve over the next six months? Is there a clear route to get there? What will they need to do or learn to get there? How can you support them as their manager? If you get some initial notes down between you, then they can continue to work on them after the review meeting and you can revisit them in your one-to-ones.

For example, you could prompt the conversation by asking:

- What's one thing at the company that you would love to work on?
- Is there anyone here that you look up to, and what traits do they have?
- How can you increase your impact in your work and in your relationships with your colleagues?
- What could you achieve in the near future that would make you feel proud?

The questions that you can ask are plentiful. Planning for the future can be exciting.

And that's it! The day itself should be straightforward if you've done all of the preparation that we've talked about. If you've put in the work up front, the meeting should just unfold in front of you. And even if it doesn't, just keep the thought bubble over their heads, ask leading questions to provoke conversation, and listen actively and attentively. Good luck.

Your Turn: Review Yourself!

Look back over the example questions that we've outlined in this section. Go through and answer them for yourself, even if it isn't review time right now. What do you think of your answers? Were you surprised by any of them?

How to Talk About Money

When performance reviews are at the end of the year, they have another piece of pertinent information attached: salary increases. This complicates things.

When performance reviews also act as the grand unveiling of a salary increase, people tend not to engage as well in the performance discussion, which is what these meetings are *really* about. Let's imagine that both of you know that you're going to tell them their pay raise at the end of the meeting. In the lead up to the pay raise surprise, people will sit there wondering when you're going to tell them. As soon as you've told them, they're either extremely happy and begin thinking about what they are going to do with the extra money (A holiday? Overpay their mortgage?) or they'll be seething because it's not what they expected. All the while, the useful performance conversation floats on by and doesn't land.

Instead, inform people of their pay raises at another time and let them know that this is going to be the case. Don't let money distract you from a focused conversation around performance. Your company may already inform staff of their pay raises by an existing system. If they don't, and it's your duty, as their manager, to do so, then it can be helpful to discuss with your staff how they'd like to be told. Some may prefer the news to be delivered via email, whereas others may prefer face-to-face. If you have the opportunity to craft the delivery to their preferences, then do so.

When you eventually do have that conversation around pay, come prepared with justification and context for their pay raise. How are the percentages calculated? How much is it driven by their performance versus company performance? Is everyone getting the same this year, or are they a special case for some reason? We'll touch more on the subject of pay in *Dual Ladders*, where we'll discuss career tracks and compensation.

Be prepared for dissatisfaction. Always leave your door open for pushback on salary increases. No matter how hard you try, there will always be staff that aren't happy. They may have convinced themselves that they've done everything they need to get a 10% pay increase, yet the company has missed its revenue target so everyone is just getting an inflationary increase instead. Often it's entirely out of your control.

Depending on your company, you may already have a process for those that feel that their pay increase isn't fair. Perhaps no such process exists. Regardless, you should leave your proverbial door open to listen to any concerns that your staff have about their pay. Even if you can't do anything about it this year, you can use it as a catalyst for discussion about what they would like to earn in the future so that you can both work on a career progression for them to reach their income goal.

Where to Next?

So, there you have it. That's everything you need to know to give great performance reviews. But like learning an instrument, mastery comes with practice. Don't worry if you still find them difficult. We all do. But stick to the principles, be confident in yourself and your abilities, and you will do just fine. We are with you.

Here's what you've learned in this chapter:

- We looked at some common *misconceptions and myths* around performance reviews and showed why they're wrong.
- You learned how to *prepare* for your reviews and how to track all of the preparatory tasks through a tracker document.
- Then we looked at the *written review* and how to create your own review forms that both you and your staff contribute to.
- We learned how to gather *peer feedback* and gave you an email template to do so.
- We looked at how to *perform on the day* so that the review meeting is successful for both of you.
- Lastly, we touched upon how to deliver the news of *salary increases* and why you should keep this separate from the performance review itself.

Another action-packed chapter over. Next up, it's all about hiring.

If opportunity doesn't knock, build a door.

Milton Berle

CHAPTER 7

Join Us!

You're sitting in your one-to-one meeting with Lisa, your manager. As you finish summarizing your plan for rolling out the latest upgrade to the ingest pipeline, you notice the corners of her mouth beginning to form into a smile.

"There's something else that I need to talk to you about as well."

"What's that?" you reply.

"We've just reforecast the budget for the second half of the year. Commercial has had an amazing couple of quarters, and the board think that it's time to start expanding again."

"OK, but what does that mean for me?"

"You're expanding too! I've given you two more headcount that I'd like you to fill as soon as you can."

You start jotting down notes in your notebook. Two headcount. ASAP.

"What sort of seniority and budget are we working with?" you ask.

"Whatever you think you need. The board is being pretty bullish with the budget at the moment. If you can get two good seniors in, then by all means we can find the money for them."

"Nice!" you say, jotting down more notes. "So how does hiring actually work here? What do I do?"

"A good question," Lisa replies. "We like to give hiring managers—that's you—the responsibility for pretty much the whole process. Our HR team will help you put your job descriptions out there, but what you do from there is your call. Whatever works best for your team."

"OK," you say, slowly. "I'll have a think about it and have a read around."

"If you have any questions just let me know." Lisa stands up from her chair, sidesteps toward the door and opens it for you. "I'm looking forward to seeing you get some new faces!"

"Totally!" you reply, parting ways.

But how does hiring work? How do you know who you should hire and how do you describe them in a job description? How should the interviews be structured? What should you look for in a good candidate? The questions start to percolate inside your head...

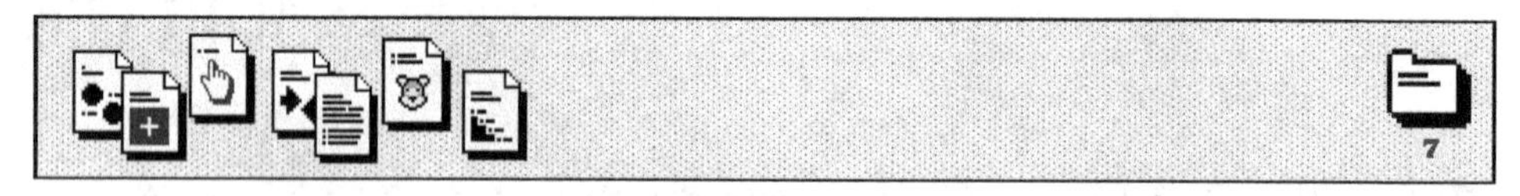

Well, look no further. This chapter is going to be all about hiring, from thinking about the people that you want to hire to having them accept your offer.

In this chapter you're going to learn how to:

- *Decide who to hire.* You'll consider the team, their skills, and their dynamics and come to a conclusion about who is best to recruit.
- *Write job descriptions.* You'll see how to describe the roles that you want candidates to apply for, and you'll do so in a way that is appealing and doesn't implicitly discriminate against certain types of people.
- *Run an interview process.* You'll consider a number of options for interview stages, from phone screens to making the offer.

It goes without saying that the hiring process in the technology industry has been fraught with issues. Searching online for "tech job interviews" or "how to do tech interviews" will present to you all sorts of conflicting information, such as advice to candidates to study whole textbooks worth of computer science problems in order to stand a chance at getting through the door at the world's best companies. You'll also find, in equal measure, horror stories about endless interviews, rejection, embarrassment, and sweating profusely in front of whiteboards.

What I just described is *not* what this chapter is going to be about. You're going to learn it's possible to find fantastic people by being kind, considerate, reasonable and *human*. Hiring doesn't have to be painful. Now that you're a manager, you have the power to be an agent of change for our industry.

So, are you ready to start adding more people to your team? Let's find out how!

Picking Who to Hire

Clearly before you start hiring you should know who exactly you want to hire. This might sound like one of the most obvious statements in the world, but there are subtleties to finding the right people to make your team even more effective. Hiring managers fall into some common traps when given the ability to hire anyone into their team:

- They just *hire more engineers*. No matter what the problem is, just throw more engineers at the problem, right?
- They just *hire the most senior person possible*. Surely the most senior candidate will have seen it all before. Forget about these people with only a few years' experience. Let's just get the experts in, surely?
- They *hire people just like themselves*. Culture fit means "one of us," doesn't it?

Wrong! There's much more to putting together a highly functional team than just finding seven carbon copies of yourself. Remember that as a manager, you are trying to find ways to increase your output, where your output is defined as:

The output of your team + The output of others that you influence

Teams are *interdependent* by nature. In technology we work on complex problems that benefit from a diverse range of skill sets, experiences, and viewpoints to produce the best possible work. Today's cross-functional team typically consists of engineers, one or more designers, one or more QA, an agile practitioner, a product manager, and an engineering manager: fully functional teams that can build software from end to end. Assuming you aren't replacing a member of staff that has departed—and even if you are—an opportunity to hire is an opportunity to pause and look at the team as a whole.

Think of the following:

- In a typical week, where does the team tend to slow down or get blocked? For example, do tasks bottleneck on QA, or are there simply not enough back-end engineers to balance the output of the front-end engineers?
- Which is the area of your team that has the least seniority? Are there dependencies outside of the team for any particular type of work where it would be advantageous to have that support inside the team instead?
- What is the team's opinion on where they are lacking support? Who would *they* add if they were choosing?

Before you rush to pull the trigger on hiring yet more engineers, step back and consider the whole interdependent team and seek their opinion on the matter. It may be the case that adding an additional QA engineer will be more effective for increasing the overall output of the team when compared to adding another JavaScript developer who adds to the throughput bottleneck. Likewise, if the team is often dealing with complex streams of work and has difficulties with prioritization and organization, then it might be more valuable to hire an agile practitioner to focus on team efficiency rather than making it even more disorganized. Don't just opt to throw engineers at the problem regardless. It's up to you to optimize for the *whole team's output.*

More Talent Isn't Always Necessary

In addition to picking the right skill sets to add to the team, you should also be looking to find someone who is the right level of seniority. "But," I hear you cry, "shouldn't we just get the most senior people possible?" That's a good question. And most certainly you should be trying to hire excellent seniors. However, this isn't always the easiest thing to do, nor is it always the best solution to your situation. A couple of reasons explain why.

Are you familiar with the classic project management constraints, typically called the *iron triangle* or *triple constraint*? It states that you can model an ideal project with three categories: *good*, *fast*, and *cheap*.

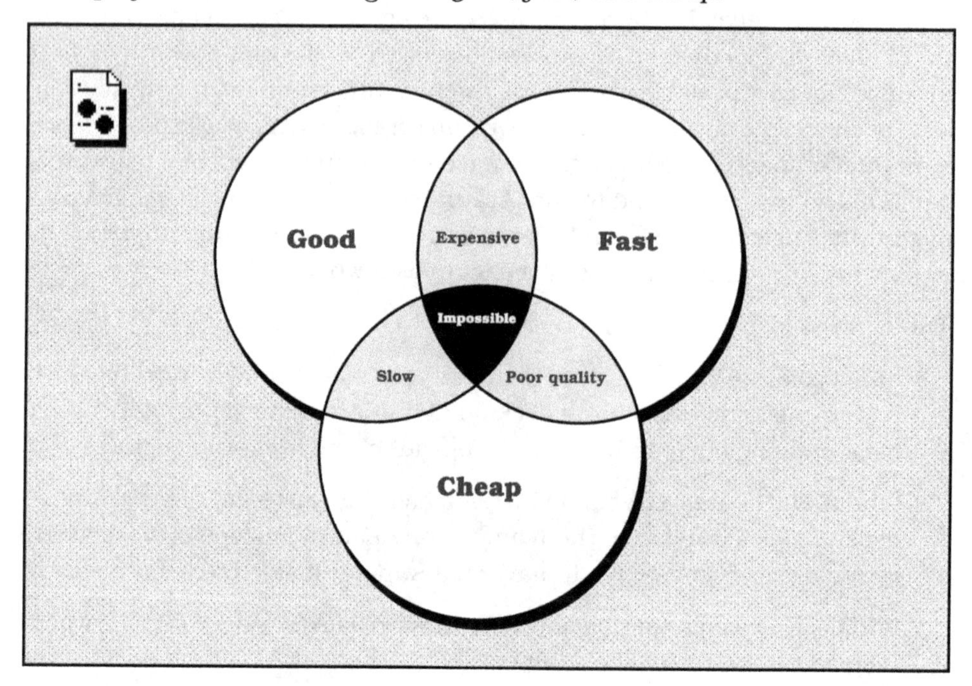

However, it states that in reality your project will typically fall into two of these categories, but not all three:

- A good and fast project will not be cheap.
- A fast and cheap project will not be good.
- A good and cheap project will not be fast.

Hiring parallels this, and we can slightly rework the iron triangle definition to capture the dilemma that you will face as a hiring manager:

- A senior hire made quickly will probably not be in budget. Typically, fast senior hires will require headhunting, which means bringing out the big bucks to coax people away from their current jobs and potentially paying expensive recruiter fees.
- A fast hire that is within budget will probably not have all of the experience that you want. You're at the whim of whoever is actively looking for jobs at the time that you put your advert out.
- A perfect senior hire that is within budget may not happen on the timescale that you want (that is, now).

Given that you have a headcount in your team to fill, presumably because the business wants you to maintain or increase your output, you can't wait forever to fill that vacancy. So as a manager you'll need to do the best that you can, within the budget that you have, within a reasonable timescale. What this means is that you'll be writing job adverts and interviewing for the *potential of candidates to grow* as much as you are assessing their current skill set. You'll see that we won't be writing job adverts that have a long laundry list of exact skills for your team. We'll be promoting an opportunity to experience something new and to grow.

Don't feel dismay that you're never going to get the perfect candidate every time. It has been shown that there can actually be such a thing as too much seniority in teams. *A study showed [SSAR14]* that in sports that have highly interdependent teams, such as soccer and basketball, there can be such a thing as too much talent! This contradicts the popular belief that there is a monotonic relationship between talent and performance; that is, the primary way to increase performance is to increase the amount of talent within a team. What the study found is that in interdependent team sports, an increase in talent increases performance to a point, after which further increases in talent become detrimental as the coordination within the team suffers.

This can happen because a winning team may have one or two stars that the rest of the team coordinates round to use them to their full effect. When assembling a team of all-stars, the natural order of the team begins to fall apart: in a team where each player is usually the center of attention in the game, how will a new pecking order form? Often it doesn't, and an all-star team can get beaten by one that is more collaborative but less individually talented.

But we're not playing sports, are we? We're building software. Yet, there are many parallels to overstaffing a team with seniors. For example, seniors may be used to pointing in a particular direction and then having the rest of the team align and support them in getting there. Too many strong opinions may cause conflict and disarray rather than getting there twice as fast. Additionally, a good balance of senior staff with those that are less experienced gives plenty of mentoring opportunity for the senior, which in turn improves the whole team in an interconnected, interdependent manner.

So perhaps the fact that you're not going to find the exact fit for your role every time doesn't matter. Instead, see hiring as an opportunity to find people that are motivated to learn, grow, and overcome challenges together. Those are the people that will elevate the whole team and improve it in the long run. Not only will you wait forever to hire three LeBron Jameses, they might actually end up falling out over who gets to take the shots.

A Note on Culture Fit

The last part of thinking about who to hire is not something you will explicitly seek, but instead is something that you should be open to. I'm sure that you've probably heard all about "culture fit" and what it means to ensure that the people you hire adhere to it. It could be argued that culture fit is impossible as a concept: culture, unless monolithic and dictatorial, is not possible to "fit" into. People instead add or contribute to a culture.

Before you start filling your mind with visualizations of the hypothetical person that you're going to find for your team, remember that you're looking for the following traits when assessing for their suitability:

- Motivation
- Willingness to learn
- Collaboration skills
- Good communication
- Empathy and kindness
- A diverse background and viewpoint
- Alignment with both the company's and your values and ethics

But *not* these:

- A background or education similar to yours
- Their sense of humor
- Someone you'd enjoy going drinking with
- Someone who has the same hobbies as you

We'll look at how to assess these traits in the sections that follow, but get it out of your mind now: *you should not be looking for someone just like you.*

So, before we move on to writing the job description, think about who you want to add to your team.

- Where is your team lacking right now?
- Does the team agree with your judgment here?
- What skill set should the new hire have?
- What sort of seniority would be desirable and acceptable for the dynamics of the whole team?
- What sort of budget is the company allocating for the role, and how does that complement or detract from the person that you're ideally after?

When you've thought about what sort of role you want to advertise for, you'll need to get a job description together. We'll do that next.

On Hiring Friends

Should you hire your friends? Of course you should, assuming you want to do so because they are good at what they do and you think they will fit in well at the company. However, you should be mindful of some things when recommending your friends for roles:

- Remember that each of us has conscious and unconscious biases when it comes to assessing candidates. You'll already be heavily biased in favor of your friend. You should make sure that they still interview for the role and that you keep yourself out of the process as much as possible, other than perhaps being present to greet them and also catch up with them at the end of the interview.
- Consider that if you are recommending a friend for a role on your team, then you're going to have to potentially deal with an awkward situation: you'll be their friend but also their

On Hiring Friends

boss. Be sure to have a frank conversation about what it really means to have that relationship at work—that you will expect them to perform—and if it looks like it might change your friendship, then maybe recommend them for a role under a different manager. Friends shouldn't be treated differently to other staff on your team. That's favoritism.

Writing Great Job Descriptions

Nobody's going to know about your role unless you write a job description, since it turns out that having your role get noticed is surprisingly hard:

- Technology talent is always in high demand, so it's likely that you're going to be competing with potentially hundreds of other companies for job seekers.
- You may only have tens of seconds to pique the interest of a potential candidate when they read your job description. Even if you feel that the company that you work at is the best place you've ever worked, it's likely they don't know much about you.
- It's easy to write a job description that repels the reader rather than attracts them, even if your intention was the opposite.

This section will help you write great job descriptions that will give you the best chance at attracting as many applications as possible. We'll show you:

- A *reusable template* that will help you write fantastic job descriptions.
- How to write in a *style* that is gender-neutral and appealing to the reader.

Once you've been through this section, you'll already be able to write better job descriptions than many companies out there. That's not a bad outcome, is it?

A Job Description Template

OK, so let's look at the rough template for job descriptions, and then we'll hone in on each section and consider how you can write them. Each job description should take the potential applicant on a journey where they'll learn:

- What your *company* is doing and why it's interesting and important.
- What the *role* is within the company that they could be doing, and what kind of work that entails.
- What you're looking for in *them* in terms of their background, skills, experience, and interpersonal traits.

- The *benefits* of working at the company, from the salary, healthcare, insurance, and whatever other perks there are to the job.
- How to *apply* with clear, unambiguous instructions.

If you're thinking that's a lot to cover in something that will probably take only several minutes to read, then you're right! Crafting an excellent job description takes real skill.

Describing Your Company

The first part of the job description should describe your company. It may be the case that your recruitment team already has a description that you can use, but you should check to see whether it could be made better. The question that this part is trying to answer is why the company exists and what its purpose is. It should grab the interest of the reader and make them think "Wow! I'd love to contribute toward that!"

Here are some examples that show the kind of company description you're after:

- At Novacorp, we make the world's most popular payroll software: it's beautiful, simple, and a joy to use. We're on a mission to revolutionize pay and benefits around the world. We are used by thousands of companies around the world and we ensure that pay and tax don't have to be taxing. We make the lives of HR and hundreds of thousands of employees a breeze.
- ACMEsoft is reducing the world's carbon emissions, one journey at a time. We want to save the planet and save car owners money while we're at it. Our software, installed in tens of thousands of vehicles worldwide, ensures that engines perform economically and safely.
- Dyncompsys is making every home a smart home. Our speakers, thermostats, screens, and cameras are making people happier, safer, and more energy efficient around the world. We want everyone to live richer lives.

Each of these company descriptions is nontechnical, understandable by pretty much anybody, and outlines a vision that hooks the applicant and allows them to immediately buy into it. If you're looking for more inspiration here, then have a look at the job descriptions for technology companies that you know. How are they writing about themselves in the opening paragraph? How can you pitch your company in the same way that immediately grabs the attention of the reader and makes them want to work with you?

Describing the Role

Assuming that you've got the reader hooked, it's time to describe the role. The most important thing to know is this—you want to sell the opportunity in terms of:

- Why it is important to the whole company.
- What it entails, using examples where possible.
- What it's like working for the company in general.

This means that you're not going to write a bullet point list of generic things that you want the candidates to perform. At an abstract level, the job of an engineer at most companies is fairly similar: they need to take requirements and produce software. So you'll need to go deeper than that. What does it really *mean* to do this role? For example, consider the following for a hypothetical role at Dyncompsys, mentioned earlier.

Example: Role Description

We're hiring a senior engineer to work on our Android mobile application. This app is the way that tens of thousands of people around the world control their smart home devices. From picking which music to listen to, to checking on their sleeping baby upstairs, your code will be making a meaningful difference to people's lives.

You'll be working in a cross-functional team that is located in our NYC office, but we have a flexible working policy for splitting your time at home too. Ideally you'll have worked on Android applications before, but if you're well-versed in Java and have a curious mind, you'll pick things up, no problem. We believe that nobody starts any role an as expert. We want to find motivated individuals who want to do their best work rather than finding someone who has done this before.

As the mobile application is the interface for people's smart homes, we want it to be an intuitive, beautiful experience. Here are some example projects that you may work on:

- Building the controls for new smart home gadgets that we create. (We just shipped a range of smart light bulbs that change color, and they're really fun to use.)

- Creating automation recipes for our users to make their lives easier. We recently created a feature to allow users to script their daily needs, including automatically turning on lights at night when they're away from home.

- Allowing experiences to be shared with loved ones. Are you checking in on the dog only to see that she's sleeping on the sofa? Let your family share the moment, instantly, in your video scrapbook.

You'll have plenty of opportunity to add your own input into our software. Engineers spend 25% of their time on their own ideas.

As you can see in the example, you'll want to write this section so that it offers plenty of food for thought for the candidate. It should describe the kind of work that they'll be doing and why it's important rather than being a wish list of technical skills. A candidate will likely not apply because you just so happen to use the exact same framework that they're using in their current job. They want to get behind the purpose of the company and the products and see how they can make a difference in the world with their skills.

Who You're Looking For

Next up, it's good to describe the type of person that you're after. Again, you're not after a bullet-point laundry list here. What sort of people do you want to hire? What are their characteristics and how do they fit into the team? Let's use another example.

Example: Who We're Looking For

Generally speaking, we're after individuals that are curious about the possibility of technology, are eager to learn, and are diligent and kind. Our teams work well because we place trust in them to succeed. We trust you to do well, and you will do so together with us.

As we're looking for a senior engineer, we expect you to help elevate the skills of the team through technical mentorship. We'd love you to show us how we can improve what we do. We also have a great graduate program that will allow you to mentor the future of our industry.

You'll probably have at least five years' experience working in the industry, but we care less about how long you've been active and more about what you've been doing. We'd love to hear about what you've worked on, how it was built, and why it was a success.

We believe in radical candor and healthy debate. We like strong opinions loosely held. Great ideas can come from anybody, no matter where they are in the company. You should be comfortable working with remote employees, which means using video calls and frequent written communication.

Again, we're after broad traits rather than specific skills. The focus is on the potential of what they can do when they're at the company—which is inviting—rather than listing specific skills, which is a barrier to entry. We'll see shortly that the way in which these sections are written can prevent candidates from applying, so keep the style loose, highlight the potential experiences that the candidate can have, and continue to reel them in.

Salary and Benefits

The important part! Don't disappoint. Where possible, you should always list the salary range that you'll expect to pay the candidate here. Don't let it be a surprise until right at the end of the interview process. If the company is afraid of putting their salaries out in the open, then why is that? Are they worried that they don't pay enough for the positions? Try and find out if you can put the salary on the job advert if the company hasn't already. It makes a difference. Also, remind the candidate of all of the other benefits of working at the company. Is there a retirement contribution? What about healthcare? Are there other benefits of working in the office worth mentioning?

Example: Salary and Benefits

We pay senior engineers anywhere between $150,000 and $250,000, depending on their experience and impact. We also offer excellent health insurance and dental coverage, a retirement plan with a substantial matching contribution, Friday afternoons off in the summer, and much more. Our office has a beautiful view over the river, and we still enjoy having our lunch together while watching the boats go by.

Try not to fill this section with periphery perks. There was a period of time where this section of job descriptions would contain little about actual compensation, but lots about table tennis, foosball, free drinks, and snacks. Those things are nice, but they're not that important compared to the financial reward and meaningful work.

How to Apply

Finish the job description with information about how you'd like the candidate to apply. Be specific about what you expect. Are you after a cover letter? Say so. Are you expecting them to provide details about previous work that they've done? Mention it. Don't let the candidate guess: make applying as straightforward as possible.

Example: How to Apply

Send your resume along with a short written application to careers@dyncompsys.com. Show us why you'd like to work here and what the company would be like if you were in it. If you've worked on any similar projects before, then we would love to hear about them. Also, if you got the job, what would you most be interested in working on first? Don't feel the need to write an essay. A handful of paragraphs combined with your resume is just fine. We are open to this role being part time and welcome candidates from diverse backgrounds.

Hopefully by reading through the preceding example you have a good idea of how to put together a job description that will really speak to candidates. You do this through writing in a light and inviting style that makes the reader want to find out more rather than putting up barriers to entry for them to climb over.

Style, Tone, and Gender-Neutrality

Evidence shows that job descriptions can be composed in such a way that it has the effect of excluding candidates at a subconscious level. This effect is especially prominent in female applicants. To prevent this, you should:

- *Not write a long list of skill requirements.* A *2014 study [San13]* at Hewlett-Packard showed that given a list of expected competencies for a role, men will on average typically apply when they meet 60% of the requirements, whereas women on average will only apply if they meet all of them.
- *Not use gender-biased language.* Choice of words and phrasing is incredibly important. *Research has shown [GFK11]* that there are masculine- and feminine-themed words, and we can implicitly encode gender bias into our job adverts without realizing. You can use an online tool[1] based on the research paper to check yours.
- *Focus on the work, not the time spent doing it.* Since candidates can be working parents or caregivers, flexibility is very important. If you have a flexible working policy, then mention it. If the role can be part time, mention it. Ensure that you describe the role in terms of what they'll be achieving rather than listing the time constraints under which they'll do it.
- *Champion diversity.* If you have a commitment to diversity (you should), which we will explore further in *The Modern Workplace*, then say so. A diverse workforce is a better workforce.

Setting Up an Interview Process

Now we're going to take a look at the interview process. Your company may already have one, or perhaps you're in the position where you're able to create your own. In this section we're going to look at the interview process, examine why it exists, and what you might want to do at different stages.

Depending on the size of your company, and the autonomy that you're being given, you may or may not be able to make changes to the interview process. However, that's not the end of the world. We'll ensure that you can approach

1. http://gender-decoder.katmatfield.com

Your Turn: Writing a Job Description

That's the learning over. It's now time for you to do it yourself. Then we'll move on to look at how to structure the interview process that kicks off when candidates have applied.

- Using what we've covered in this section, write a job description for your team.
- Run the job description through the online tool above that detects gendered language. How did it fare?
- Read the existing job descriptions on your company website. What do you think of them? Would you change them?
- Are you happy with the way that your company describes itself? If you think it could be improved, who can you talk to at the company to see if they're open to making it better?

interviewing confidently in a way that makes the experience comfortable and productive for both you and your candidates.

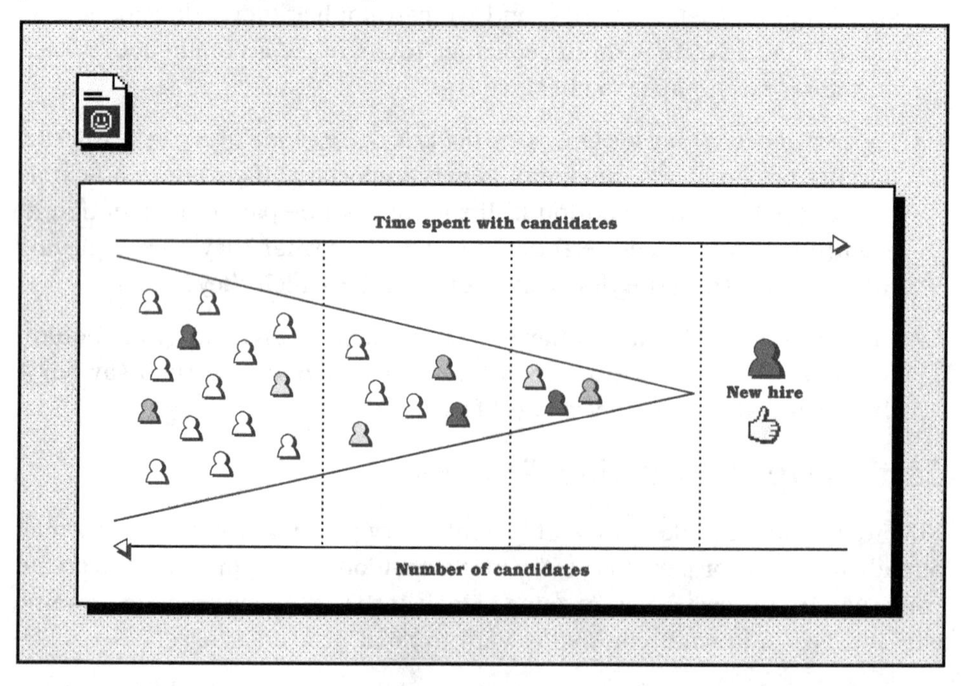

The interview process is like a funnel. At one end, you should (hopefully) have people applying for the position. At the other end, you have your chosen candidate that will accept your job offer. As candidates progress down the funnel, the probability of them being offered the job should increase. This also means that as they progress, you and your colleagues will be spending increasingly more time with them. This is why you need to iteratively narrow down your choice as they progress through the stages.

Regardless of where candidates get to in your interview pipeline, there is one important rule: each and every one of them should have a good experience. An interview process says a lot about a company. Applying for jobs is stressful. Not everyone is applying because they fancy a change. Sometimes people are applying for jobs because they've just been laid off elsewhere and need to support their family. Perhaps they have been headhunted by a recruiter and it's now your duty to sell them the role. You owe all of your candidates a professional and considerate experience with clear and timely communication along the way. In an ideal world, everybody that interviews for your company should be left with the impression that it's a great place to work. If they don't end up getting the role, they should understand why. They should be encouraged to apply again in the future if a better-suited role is advertised. Be mindful of the experience of the candidate at each step.

You'll want to design your interview stages so that you spend the least time on the people that you're least likely to hire and the most time on those that you're most likely to. Most commonly this is implemented as a series of interview stages that increase the amount of commitment from both the candidate and the company. Although the experience may differ at the extreme ends of the company-size scale—a startup interview process may be short and informal, and a large company may be long-winded and formal—a process that works well looks like this:

1. Review applications to find candidates that look promising.
2. Perform initial screening calls with promising candidates.
3. Hold a first interview for both sides to get to know each other better, typically with the hiring manager and one of their future teammates.
4. Perform an optional technical exercise, either in-person or as a takeaway test.
5. Hold a final interview with two more of their future colleagues.
6. Make an offer.

Let's have a look at each of these interview stages in turn.

Reviewing Applications

For those that are applying, you'll need to review the applications to see whether they are the right fit for the role. If you work for a large company, then you may have a recruitment team that does the initial application screening for you. This is very much an exercise in sense checking rather than a deep analysis of a candidate. Scanning an application, you should be looking for:

- Whether they have the right amount of experience that you're looking for in that particular role.
- Whether they have used the technology that you are working with, or if they haven't, whether they've used something similar enough to enable them to be productive.
- Whether they've put time and effort into their application. Although you're not marking anyone on their spelling, punctuation, and grammar, those that take time to craft their application are worth noting.
- Whether they're located near the office for non-remote positions. Are they already a commutable distance, or will they be relocating? Would they need a visa to work at the company?

Fundamentally, does this candidate look like they could be a good fit for the job? Try not to form biased judgments of candidates. For example, if they have some gaps in their resume, don't assume that they were out of work because they weren't good enough. I once interviewed someone who had a two-year gap in their resume, and it was because they were in a band that toured the world! Additionally, don't assume that because someone has moved jobs a lot that they're only interested in staying at yours for a short time. There may be good reasons for it (for example, they were a series of startups that didn't take off), and you can talk about it in the interview.

Be aware of your own biases. If you find yourself being drawn to particular resumes and you don't understand why, enlist the help of a partner to help you review. Additionally, don't discount candidates that have had an extremely diverse and varied experience. If someone has been a dancer, then a mechanic, then a musician, and is now a programmer, then isn't that a sign that they are capable of learning a multitude of new skills?

Sift the resumes into those that make the cut and those that don't. Pass those that do on to the next phase. Those that don't pass deserve a reply, typically via email. This is something that your recruitment team may also do for you if you're at a larger company. However, if you are writing the rejection letters, then remember that you want the candidate to have been handled sensitively and kindly so that they might apply to another position in the future. Be succinct with your message and be respectful in how you write it. Don't be too specific about why or mention how they compare to other candidates, and don't make empty promises about the future. Note that some companies don't even send out rejection letters. If you are sending them yourself, then you have the opportunity to do them well, thus ensuring that candidates are left with a positive impression of you and your company.

Example: Rejection Letter

Dear Steve,

Thank you for your application to the senior android engineer role. We're sorry to inform you that you haven't progressed to the next stage of the interview process. We loved your application, but we are looking for people that have more experience in industry.

We are growing, so keep checking our website for other roles as the year goes by. We'd love for you to apply again.

For those that have passed this initial stage, you should have agreed they seem to have the credentials you want for the role, and you may also have a number of questions to ask at the next stage (for example, whether they are considering relocating, given that they are a long way away from the office).

Performing Screening Calls

Those that make the initial cut should move on to screening calls. Again, depending on the size of your company, the recruitment team may also do these. This is the second of the screening stages that are meant to form an initial connection with the candidate and also ask any questions that arose during the resume-sifting stage.

It's good to use videoconferencing software such as Zoom or Skype to do these calls so that you can see each other as well as hear each other. Body language makes communication better. Here's a basic sequence of themes and questions to run through during the phone screen. After you're on the call:

- Introduce yourself and tell them how long the call is going to take. Usually fifteen to thirty minutes is enough.
- Tell them some more about the company and the role. What sort of work are you doing, and where do they fit in?

- Figure out any administrative items, such as whether they need to relocate, whether they need a visa, or whether there are any time-related issues: for example, the length of their notice period if they are currently employed and when they could start.
- Ask them about their last role and what sort of responsibilities they had.
- Ask why they decided to apply for this role, and what specifically is most interesting to them about it.
- Ask them what their salary expectations are.
- Ask them if they have any questions for you at this stage.
- Then, conclude by telling them what happens next.

Hopefully there shouldn't be that many people that looked promising at the resume-screening stage that turn out to be the opposite after an initial call. But it can happen occasionally. Also, don't throw out any obtuse brain-teaser questions at this stage. Whether a candidate can tell you how to estimate the number of streetlights in the United Kingdom or tell you how they would get out of a blender if they were only 2 centimeters tall doesn't say much about their ability to be a diligent worker. Those that pass this stage get a first interview.

The First Interview

This stage is where you begin to spend some real time with the candidate. If the candidate is located within commuting distance to your office and you are not a remote company, then it's good to get them to come in person, since it's easier to make a more human connection that way. A first interview can be done in an hour, and it should ideally consist of the hiring manager (that's you!) and somebody else who will be working with them on the team. Don't have more than two people in there interviewing—it's intimidating otherwise.

Before the candidate is invited, make it clear what they should expect from the interview so they feel as comfortable as possible. Tell them what to wear (candidates turning up in a suit to a casual-clothes company will make them feel embarrassed) and also roughly what you're going to be going through. For example, you could tell them that:

- The interview will be one hour.
- It will be with the hiring manager and one other person who would be on their team.

- It will consist of some conversation for each party to get to know the other and to find out more about their past roles and experience.
- There will be some technical exercises that will be done collaboratively.
- There will be plenty of opportunities for them to ask questions about the company, the team, and the role.

The more transparent you can be, the more you put the candidate at ease and the more you're implicitly saying to the candidate that you want them to have the best opportunity to succeed.

Preparation

Once the candidate has been booked in to the interview, do your preparation. Both you and the other interviewer should take the time to read the candidate's application with the following questions in mind:

- What are the candidate's main strengths?
- How many years of experience do they have and where have they gained them?
- Do they have any unique skills and knowledge that will be valuable to the team?
- Have they worked anywhere that may have solved similar problems to the ones that you work on? For example, are they also working at a SaaS company on their data ingest and storage, or were they working on something entirely different?
- Do they have experience in working in a similar manner to you (for example, a cross-functional agile team)?

Think about the sorts of questions that you want to ask them in the interview after reviewing their resume. Is there anything that you have any concerns about, such as them having a decade of experience in a formal corporate company when you work at a messy startup (or vice versa)? Note them down and share your thoughts with your fellow interviewer.

Being a Great Interviewer

By coming this far in the book, you'll have already learned and put into practice all of the necessary skills that you need to be able to interview well. The communication skills in *Interfacing with Humans* cover everything that you need: balancing your spoken and nonverbal communication, managing your energy, listening more than you speak, and remembering that the interview is primarily about the candidate, not yourself.

However, bear in mind a number of interview-specific tips. Let's go through them.

- *Prepare!* Read their application and highlight parts that are interesting to you. Make notes of questions about it that you can ask. Have it in front of you while you talk to them so you can refer to it. Show that you've taken the time to read and digest the application that they spent time putting together for you.
- *Make the candidate comfortable.* Little gestures go a long way. Greet them at reception and walk with them to the room rather than sitting in there waiting for them to enter. Offer them a drink. Put them at ease by asking about how their day has been or how their journey was.
- *Be yourself.* The most important thing, as with life, is to be yourself. You don't need to be a hyperinflated version of yourself, or more serious than you usually are, just be you. After all, this person is going to be reporting to you and it's your duty for them to see as much of the real picture of their future manager as possible. Accepting a job offer is a two-way process; they need to choose you as well.
- *Describe what you do before you do it.* Given that interviews are stressful, outline what you're going to be doing before you do it. This applies to the beginning of the interview where you can describe the different parts of the interview that are coming up, but also at each individual stage. Over-describe what they should expect. For example, you could say: "We're going to do an exercise on the whiteboard now. There's no right or wrong answer and we're going to try and figure it out together." This is much less scary than handing them a pen and asking them to stand up there before they know what they have to do.
- *Allow for nerves.* Remember that there's a likely chance that your candidates are feeling nervous, especially those that are younger or less experienced with interviewing for technical roles. Being warm and kind goes a long way here. However, since your candidate is probably not thinking as straight as they would if they were sitting at their computer listening to their favorite music while programming, allow for mistakes, don't fuss over small incorrect details that contribute to fluster, and help out when necessary with examples and additional explanation.

- *Detect silence and intervene.* If you have asked a candidate a question and they're struggling in silence, then intervene. Ask whether they need more explanation or would like to work it through with you together, out loud. If they're getting really stuck, divert and ask an alternative question. Be flexible with your approach.
- *Invite questions and conversation.* The best interviews are transactional two-way exchanges. Make this clear by asking lots of questions and keeping the conversation flowing. Invite opportunities for them to teach you something ("Tell me about how you built that system; it sounds really interesting!") so that they can continue to build their confidence during the interview process via contribution.
- *Make notes.* If you're doing a lot of interviews over the course of a day or a week then it's easy to forget the specifics of your conversations with each of the candidates. Jot down reminders to yourself throughout the interviews so that you can give objective feedback without having to rely on your memory.
- *Be aware of your own unconscious biases.* We all like people who are like us. Unconscious bias is natural and unintended. As a hiring manager you need to build a body of evidence that supports your hiring decisions. Note down the answers that candidates give to questions rather than your own judgments. Don't rush your decisions. ("They were amazing! Let's hire them now!") Instead, put space between interviews and decisions and consult with your fellow interviewers at each step.
- *Find other veteran interviewers and ask for advice.* Don't trust this book alone. Go and find other people that you know and ask for their tips on interviewing. What do they advise?

The Interview

The purpose of the first interview isn't meant to be a thorough technical grilling of a candidate. After all, an interview doesn't provide the natural conditions under which engineers do their work day-to-day. Instead, this first interview is about finding out whether the candidate is someone that you can see you and your team working with and whether they are the same as (or even better than) they presented themselves on the application.

It's useful to have a running order and set of questions prepared for the interview. This reduces bias by ensuring all candidates face the same interview structure. It may be the case that the conversation will flow naturally in a way that doesn't require you to refer to them at all; but it's useful to have them to hand. For example, you could organize something that looks like the following, divided into sections, that will help you provide consistency across the interviews for each candidate.

Example: First Interview Questions

These should be a mixture of general questions that you would ask any candidate, combined with questions that are specific to the advertised role.

Introduction:

- Talk to them about the company.
- Describe what they will be doing within the team.

Their current job:

- What are/have they been working on in their previous role?
- What is the size of their team, department, and company?
- What was/is a typical day like for them in their previous role?
- How does software get defined, coded, and shipped?
- If they're still working there, what makes them want to leave?

Technical expertise and knowledge:

- How is their current codebase or architecture structured? What's good and bad about it?
- How does software get tested and whose responsibility is it to test?
- What's their favorite framework/language/library and why?
- Do they have any experience with the frameworks/languages/libraries that you use?

General technology questions:

- Which IDE do they use and what do they like about it?
- What's the most exciting piece of software that they've used recently and why?
- When did they get their first computer?

Wrapping up:

- Ask them if they have any questions for you.
- Let them know about the next steps in the interview process.

Now, let's talk about *technical exercises* in interviews. If done right, they have their place. If done wrong, they can cause major stress and panic. Here are some do's and don'ts for technical exercises in interviews.

Do:

- Work through a problem *together* with them. Don't just sit there in silence while they struggle. Imagine that you're both on the same team doing some work where you're mentoring them. If they get stuck, ask questions. Some candidates have hundreds of interviews under their belt, whereas others may have only a handful, even if they are senior.
- Pick problems that have multiple ways to get to the correct answers. Ideally pick ones that are straightforward to get some code written, but the interesting part is optimizing it together.
- Respect that they're nervous and not solving in their natural environment. Be kind.

Don't:

- Create an examination atmosphere. Talk. Think and act as if the problem is interesting for you to solve.
- Give them ridiculous brain teasers to do. Just because you may have a perfect solution in your head involving some unholy runtime casting and bit-shifting, don't ever assume that's going to come to a candidate under pressure.
- Assume that everyone is a math whiz. If your problem is to code something that is an exercise for university-level mathematics students, unless they're going to be doing this on the job, trying to making yourself look smart will make them not want to work for you. The same applies for arcane memory-management tricks, especially if that's nothing like what they're doing in the day job.
- Assume that what they're writing has to compile. Free yourself from these constraints and just use pseudocode. Their thinking process is what you want to test, not their memorization of a language and its standard APIs.

Another option is to keep away from code and do a technical exercise that involves some kind of system design. For example, on the whiteboard you could draw two stick figures with cell phones and ask the candidate if you could design a chat application together. Where would they start? What language would they write it in? What would need to be done if the history of the conversations needed to be viewable? Where would it be stored and how?

With exercises like this you can follow along with the candidate and use your intuition to center in on areas where the candidate is most comfortable. When you ask them how the history should be stored and you can see that they're experienced with relational databases, why not start pondering the schema with them together? They'll get to demonstrate what they're good at, and you can have some fun together while you do it.

Now that you've got your rough outline of the first interview in place, you can go ahead and do it. Think of it like a one-to-one with one of your staff. Be natural, open, warm, and kind. Believe that each candidate could potentially be your next colleague, and believe in them rather than confronting them. Remember that even if they aren't right for the job, you want them to have such a good experience that they wouldn't hesitate to apply again in the future.

At the end of the first interview you and your colleague should have a good idea as to whether they should progress to the next stage or not. If you're on the fence, side with a no. Only progress candidates through a stage if you're absolutely sure—so far—that they're a likely hire. If they don't proceed, ensure that you provide a prompt rejection letter that is succinct and respectful, since the candidate has given up their time for you.

Optional Technical Exercise

After the first interview, there's the option of giving them a technical exercise to take away and do at home and then submit it when they're done. These take-home tests have been controversial, but often they can be useful. The reason that they're optional is that sometimes it's already evident that the person can program well. They may have a solid resume full of senior positions at known and respected companies, or they may have a GitHub full of work that they're actively contributing to. In this case, a take-home test probably won't teach you any more about their ability to program, and in fact it might even make them want to work for you *less*!

However, for many people, it's hard to know how good their programming skills are. A take-home test is an opportunity for them to do some work in an environment that they are comfortable with, without all of the pressure of interviewers looking over their shoulder. This is especially useful for nervous candidates and for graduates: they get the best chance of doing a good job.

You need to ensure, however, that your take-home test meets the following requirements:

- That the scope is extremely clear. One way of doing this is providing a repository for a project that already performs some functionality and asking them to extend it, or to fill in methods and classes that have intentionally been left blank. Don't leave them to set everything up themselves.

- That the time that it takes to do the exercise is short, say, a couple of hours. Anything longer than that and you're doing two bad things: you're not respecting their free time (people have families) and you're making them work for free (that's not cool).
- That the task in some way, as much as you can, mirrors something that they'll be doing in the day job. If you're a company that uses social media data, why not set a project up that lets them work on a data set of tweets? If you're an aeronautical engineering company, why not have them fix some bugs in a toy version of the software that controls airplanes?

Your take-home test, like the interview, should make them feel like you've designed it with them and their experience in mind. Put the work in—it's worth it for your reputation as an excellent interviewer.

Final Interview

Assuming your candidate has made it to the final interview, you should have a strong feeling that you'd like to hire them. The final interview is primarily about:

- Spending more time getting to know them.
- Discussing their take-home test to understand more about how they tackled the problem (assuming they did one).
- Letting others interview them to ensure the opinion of the initial interviewers is cross-referenced, valid, and less subject to your own biases.
- Clearing up any questions you had in the previous interview.
- Depending on the size of your company, meeting the CTO or CEO, in case they want to pop in and say hello. Do let your candidate know if this is part of the plan—it can be intimidating!

One option for you to not overcrowd the candidate with too many interviewers is to begin with yourself and two others from your team, have everyone introduce each other, kick off the discussion, and then leave the room until close to the end. You can make it more of a two-way conversation this way: their future team mates can get to know them and perhaps do another technical exercise together, and both sides can ask each other questions. Also, think about seating. Interviewing by panel is intimidating: think about less confrontational seating arrangements.

Ensure that the candidate gets to use this interview to find out what it's really like to work for the company from their future peers. This builds trust between both parties as well as getting the interview done. At the end of the interview come back in and ask how they got on, and allow them plenty of time to ask as many questions as they like about what it's like to work at the company.

After the interview is over, between you and three others, you should have a good idea of whether you'd like to make them an offer or not. Prepare your judgments in isolation. Don't confer, as it introduces bias. If all of your interviewers say yes, go for it. If you're collectively on the fence, it's best not to proceed. You owe it to your team to hire the best people that you can. Only make offers to those that you are sure will do a great job, otherwise your team will be the ones that have to potentially support an underperformer, and they're not going to like you for that.

Making the Offer

At this point, you should have already done your research and have a salary in mind. However, before you make an offer to a candidate, you should also consider the following:

- The salaries of their peers at the company, so that you can make sure that they're paid as fairly as possible compared to their peers.
- What other companies in your area might be offering for similar roles (this may involve some research).

Your offer needs to try and satiate these three categories, although doing that in practice can be difficult. Often, you'll need to overpay someone compared to their peers to get them in the door, and then it's your responsibility as a manager to make sure their peers are able to square out with them over time.

Your job at this stage is not to try and scrimp and save. Don't lowball candidates. It's insulting. You're not here to play a game with them where you start low and then they incrementally barter with you for higher compensation. You should be able to offer them a salary along with justification as to why you're offering that. If you were advertising a salary band on the advert and you're offering the top of that band, then say that's because you see them toward the more senior end of that role. If you're offering toward the lower side of the band, it should be because their experience isn't quite there yet, but they will have plenty of opportunity to increase their compensation as they continue to perform well.

As you have seen, to get to this stage, a fair amount of time has been invested in the candidate. That means it can be incredibly frustrating if they decline your offer. Perhaps they've accepted an offer elsewhere already and you weren't top of their list, or perhaps you were but some other company is going to pay them an amount of money that's disproportionate to their experience. These things happen. Let it go and move on. The right person will turn up, you'll see.

From Hiring To...

Wow, we made it! As you can probably tell, hiring is quite involved. However, it's absolutely worth it. Here's what you've learned in this chapter:

- *How to work out who to hire into your team.* You considered your team, their output, and then worked out who best would increase it.
- *How to write job descriptions.* Not only did you learn how to put them together, you also understood how job descriptions can encode bias and unintentionally discriminate against candidates.
- *How to run an interview process.* From phone screens to the final interview, you've been through it and can replicate the same thing in your own company.

Even though the hiring process requires a lot of effort and organization, getting great people into your team pays dividends in the future. Doing so in a way that treats all candidates with respect does wonders for your reputation as a company and as a hiring manager.

Conversely, employing the wrong person because you haven't put the effort into your hiring process will make your life a misery. Since a bad hire will be a member of your staff, you're going to have to deal with the fallout. This distracts you from what you really want to do, which is building great technology. However, people are always going to leave the company, either voluntarily or because they're being forced to leave. We'll learn all about that in the next chapter.

Thank you, Mario, but our princess is in another castle!

> *Toad*

CHAPTER 8

Game Over

You're typing away at an email when you notice a figure approaching to your left out of the corner of your eye. You finish your sentence with a full stop and hit send, at the same time swiveling your chair to face your visitor. It's Sam.

"What's up?"

"Have you got a minute?" she replies.

"Sure thing. Shoot!"

"Can we grab a room?"

You nod and stand up, looking around at the meeting-room doors.

"Fermat's free," you say while pointing to the corner of the office. You both begin walking over there.

"Is everything OK?" you ask.

"Yeah, just something I wanted to talk to you about," Sam replies.

She enters the room and you follow. You close the door behind you and switch on the light. You notice that Sam hasn't taken the seat at the table. Her body language has become more awkward. There's an uncomfortable silence. Sam breaks it.

"I'm handing in my notice. I'm hoping to leave at the end of this month."

You feel your cheeks flushing red.

"You're leaving? What's going on?"

"As of this morning I've accepted an offer elsewhere, so I'm looking to wrap everything up in the next four weeks so I can start there on the first of next month."

You didn't see this coming. You're not entirely sure what to say. You fumble around your brain for some words.

"I'm...sorry to hear that. Let me know if there's anything you need while you're still here."

Sam nods. "Will do," she says. She smiles and exits the room.

You feel instantly stupid. Why didn't you ask where she's going? Why's she going? Was it the money? Was it the project? And when did she even have interviews at other companies when she's been here every day for the last few months? And aren't you meant to now go and give her a counteroffer to try to keep her at the company? Oh, but you didn't even ask what the new job was and what it was paying!

You sigh and lean against the door. What should you do next?

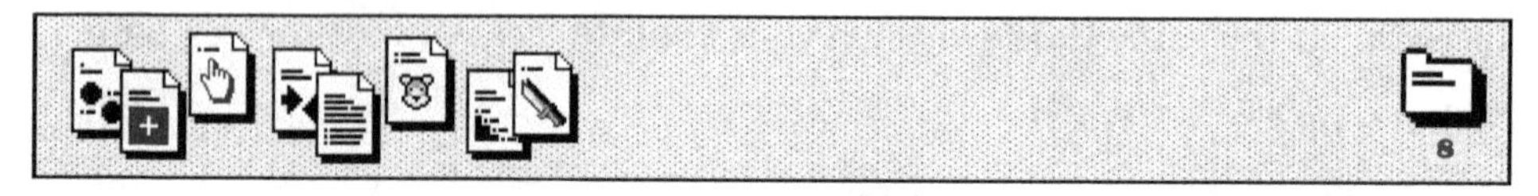

If you're a manager, then this is going to happen to you. People will always leave. Sometimes you'll see it coming, sometimes you won't. Sometimes you'll want them to leave. Sometimes you'll make them leave. Either way, you're going to need to get comfortable with the fact that there will be a natural turnover within your team that you'll never be able to win against.

In this chapter we're going to be exploring all of the wonderful ways in which people are going to leave your team. And they will, believe me. You're going to learn the following:

- *That people will always leave.* It's a fact of life. We'll look at studies that show what sort of turnover you should expect and how it can actually be beneficial.
- *How to know when people should go with your blessing.* Sometimes people will leave for good reasons. What are they, and how should you handle them?
- *How to fight for staff to stay.* Your best staff will be hot property on the job market and a highly paid job offer is only a LinkedIn message away. How can you try to prevent staff from leaving, and is it possible to predict?
- *How to make people leave.* Lastly, there will always be people that are bad performers that you'll need to address. We'll look at a common way of addressing performance issues: Performance Improvement Plans.

Are you ready to tackle the hard stuff? Let's get going.

People Leaving Is Normal

Firstly, let's dispel a myth: it's completely normal for people to leave your team, your department, and your company. In 2018, based on LinkedIn data, *it was reported [Boo18]* that the tech sector in the United States has a yearly turnover rate of 13.2%, the highest of any other business sector. Yes, that's even higher than the retail sector, which was second highest at 13%. Think about that for a second: if your department has 1,000 staff, by the end of the year, on average, you'll be down to 868 staff through attrition alone. That's a lot of hiring to do to keep the department the same size. As you saw in *Join Us!*, hiring people is hard work, so if people leave, you want to make sure that it's for the right reasons.

Now, my parents would certainly be shocked by that statistic. My father's last job before he retired had him doing eighteen years of service. In this new world where everyone's tenure is becoming ever shorter, how can we hope to make the best of the situation? How can we adjust our own expectations and feelings about people leaving to make the situation as painless as possible?

People are always going to leave. It's normal and it sucks. Just read it a few times and let it sink in. As a manager, you're doomed to failure if you think that you're going to keep everyone in your current team indefinitely. You will only set your expectations such that you'll feel terrible when someone does hand in their notice. Be comfortable with the fact that all of our careers are different and we're all motivated by varying aspects: challenge, location, comfort, working hours, programming languages and frameworks, friends, and new opportunities to name but a few. They can all be conflicting forces in career decisions.

However, believe it or not, *turnover can be good for you [Ash19]*. Companies change all the time, especially if they're small and growing quickly. The people that were hired at one point in the company's history may not be best suited for the company at its next stage of evolution. A natural turnover ensures that you have more opportunities to hire people who are best suited to the particular challenges that you're facing at that time.

You can split employee departures into two camps:

- *Voluntary*. They're leaving of their own accord.
- *Involuntary*. You're making them leave, typically for performance reasons.

Sometimes these categories are referred to as *regrettable* and *non-regrettable*. There are clearly two things to focus on here: making sure that those that are leaving voluntarily are doing so for positive reasons and that there's effectively nothing you could—or should—want to do to make them stay. We'll look into voluntary departures in the first half of this chapter.

The second reason—involuntary departure—is where you're having to terminate employment for some reason, typically because of poor performance. It's your responsibility as a manager to ensure that employees that are unable to be transformed into good performers don't stay in your organization. After all, your bar is only as high as your worst performing employee. Hiring well gets you most of the way to ensuring that this won't happen, but people and situations change. We'll give you the tools you need to ensure that you can manage poor performance and turn it into good performance, or in the worst case, arrange for that member of staff to depart.

We'll begin by looking at voluntary departures.

When Staff Leave

Our lives and careers are more connected, varied, and challenging than they have ever been, and this is especially true in the technology industry. Those at the start of their career have more societal pressure to get the best experience they can get, no matter where it is. I see engineers bouncing between companies and cities year after year to keep progressing along their desired life and career path. You can't hold people back.

Likewise, in an economy where house prices are continually rising, having just one person in a relationship working full time is much less likely, especially if both parties are in similarly competitive industries. There exists the *two-body problem [Rog12]* for academics, which highlights the struggle faced by a couple when each is trying to find tenured academic work where they can still live together as a family. For technology and creative jobs, we're not limited as much by the lack of opportunities, but couples can face the tension between being able to do the job they really want and the location they want to live in.

Our career drive often throws a grenade under our natural human instinct to settle. As such, as a manager, you need to get comfortable with people leaving to progress their lives and careers.

Good Reasons for Leaving

Those that you are managing are always assessing which other opportunities are out there. When faced with someone telling you that they are going to leave, you can be quick to get angry and think that your employee is giving you another problem to deal with, that they are ungrateful of their position, that they're just chasing money or prestige, or that they're taking the easy way out of a hard year at the company. This is rarely the case. People leave for many legitimate reasons that show no malice toward you as their manager. For example:

- *New opportunities.* Sometimes there's no room for an employee to be promoted any higher in your department, so instead they're going to find that role elsewhere, or they wish to join a company where there's more room for that role to be created, such as an early-stage startup in a phase of fast growth. Additionally, they may have worked at your company for a long time and just fancy a change in surroundings and the type of work that they're doing. Maybe an opportunity has come up to work with their best friends. That's totally natural, and it's not your fault.
- *Family.* Their partner may have been offered a job of a lifetime elsewhere and they need to relocate and find a new job nearby. They or their partner may have aging or sick parents and need to leave to offer the right level of care, especially if their family aren't local. They may want their kids to go to a particular school, perhaps because their child needs a particular education, either through learning or physical disabilities or maybe through academic brilliance. You can't control these things, so just let them be.
- *Compensation.* Sometimes your staff are in the right industry at the right time and get offered a life-changing compensation package elsewhere: the sort of package that could mean they could retire ten years earlier or that their partner could quit their job, take a year off, and then start their own business. That's just the nature of a free-market economy, and as hard as it is for you and the team, just be happy for them. It's a nice thing.

In these situations, you've not done anything wrong. A swirling and complex web of life outside of work pulls and pushes people in a multitude of directions. When somebody hands in their notice to you in order to depart for good reasons, your main aim should be to make their exit as amicable as possible. You can do this by being the facilitator in their departure. You can do the following to ensure that it goes as smoothly as possible:

- *Work out an end date that works for both of you.* A diligent staff member will want to make sure they exit on good terms, so talk to them about what they're currently working on, how it affects the team, and when might be the best time for them to go. If they've already agreed to a date with their future employer then so be it, but, regardless, jointly put together a plan about what you'd both like done between now and their departure date.

- *Ask whether they'd like a reference.* They might not need one, but it's common courtesy to ask, even if it's just on their LinkedIn profile. It shows the outside world that this indeed was a good leaver and you both cooperated while they were on their way out.
- *Focus on what needs handing over.* Do a deep dive into everything that they're working on. What needs handing over to other members of the team, and how? Would it be worthwhile them spending a week thoroughly documenting the work that they'd been doing recently, or is it worth them scheduling a session with the team to run them through it?
- *Think of the ideal replacement and get hiring.* Now that you're going to have to replace this staff member, think of the team's output and think who will best replace them. Then get that job description written and advertised. Refer to *Join Us!* to see how to do this.
- *Let them go with your blessing.* Most of all, be kind and appreciative of the time that they've spent contributing to your team. Tell them that if the new gig doesn't work out then they're always welcome to come back into their old role, no questions asked. Give them a much-needed safety net while they embark on a new journey. You'll be surprised how many *do* come back if you're a good manager.

Your Turn: Good Leavers

Before we move on, a couple of questions for you:

- Think back over the past few years. Who have you worked with that has left their role for good reasons, and how did their manager and team react to the news? Would you have done anything differently if you were their manager?
- What about your own career history? How many departures were for good reasons? Is that how you ended up in your current role?

Bad Reasons for Leaving

Sometimes, people leave for bad reasons. But what do we mean by bad reasons? I'm not talking about a situation where they've stolen the office toaster or put a dead fish in the air conditioning. Typically these are the departures where you, as their manager, are totally surprised that they are going—where you are caught completely off-guard by someone handing in their notice, insofar that you know in retrospect you could have prevented it from happening. I've heard these situations called "zingers." Often they have a similar root cause: a lack of open and honest communication from both parties, which results in simmering issues not being caught early.

Some examples of these zingers are below.

- *Compensation.* Your direct report was unhappy with their end of year pay raise, yet they felt that they were unable to talk about it openly with you. They got continually more annoyed about it to the point that they answered that email from a headhunter and went for an interview elsewhere. You found out about this for the first time when they had accepted the other job offer, giving you no opportunity to try and rectify this pay issue yourself.
- *Issues with coworkers.* Your direct report simply couldn't stand one of the people on their team, and every day over the last six months has been immensely frustrating for them. They don't have any issues with their coworker's work; in fact, it's very good. However, their personalities clash badly and they didn't want to raise it to you as they felt it was a personal issue that would reflect badly on *them* rather than a professional one that could be resolved. It got so bad they applied elsewhere.
- *Career progression.* Your direct report handed in their notice because they've been offered a role at another company that's a significant level up on the career ladder. They cite that there were no opportunities for promotion in the department. However, you know that in a few months a new team will be created and they would have been a perfect fit. But you didn't even know they were interested in being a team lead! Argh!
- *Lack of challenge or new experiences.* Your direct report has become exceptionally bored of writing code for the API and would love to increase their skill on your data ingest architecture instead. They didn't feel like they could ask to change teams, as they felt that they were employed for the role that they are currently doing. You never suggested anything else because they seemed to be so diligent at what they were working on! You, however, know that they could have just asked to change teams. Why didn't they say anything?

So, what can you do to prevent this? Hopefully you'll see a common theme at play here: a lack of an open, transparent, and candid relationship between yourself and your staff member. Fortunately, by reading this book up until this point, you'll have already learned much more than many other managers about having strong relationships with your staff. This makes it much less likely to happen to you; however, it still can happen.

What you need to ensure is that you're building conversations into your one-to-ones regularly about:

- *Career progression.* How often are you talking to your staff about their careers? Where do they want to go in the next six months, two years, and beyond? You've already learned how to give performance reviews that allow for introspection and goal setting in *The Most Wonderful Time of the Year.* However, later in *The Crystal Ball* you will learn how to do a two-part career-planning exercise that will allow you to dig deep into the desires of your staff. You can refer back to this exercise continually over the years.

- *Maslow's hierarchy of needs.* Have you talked about the model that you learned in *The Right Job for the Person* with your staff yet? You should. You should also revisit it occasionally to see whether your staff are working toward self-actualization or whether there are issues further down the pyramid that are nagging at them. If so, what can you do? Can you fix them?

- *Pet peeves.* Sometimes the tiniest annoyances can happen over and over again until they make you explode with rage. Is there anything that repeatedly bugs your staff? Do they have frequently frustrating interactions with anyone on their team? Are they secretly fuming at the state of technical debt in the codebase? Are they tearing their hair out over your ancient build system? It's your job as a manager to uncover these frustrations and turn them into opportunities for your staff to make them better. Facilitate that difficult conversation with their colleague. Let them fix that nagging technical debt. Let them propose a better build system to replace your current one and find them the time to do it. There's nothing more satisfying than scratching that itch that you previously couldn't reach.

Your Turn: Bad Leavers

A couple of questions for you again:

- Cast your mind back again. Did you work with any colleagues that left for bad reasons? How did the team react and what did their manager do about it?

- Again, what about your own career history? How many of your own departures were for bad reasons and why?

Most of all, if you truly care—which I'm sure that you do—your staff are more likely to be open with you. Be interested in your team's life outside work, in their emotions and their hopes, both for their life and their career. Many clues will surface that you can use to keep your staff happy. You might just prevent people from leaving.

Fighting the Good Fight

Regardless of whether a member of staff has handed in their notice for the aforementioned good or bad reasons, the act of handing in their notice puts the ball firmly in your court. What are you going to do next? This is a conundrum that faces many managers. At the most abstract level, it boils down to two choices:

- You accept the situation and initiate the process for them to leave.
- You fight for them to stay.

No battle is fought without effort or concession, so you'll need to think long and hard about what you want to do next. But there's an added element of difficulty. Typically, to receive a job offer from another company, your staff member will have committed a sizable amount of time, energy, and emotion into deciding to interview elsewhere, research open positions, put together applications, do multiple interview rounds, and so on. Regardless of the fact that average tenure is shortening in technology, changing jobs is still something that takes considerable effort.

Although some people do regularly interview elsewhere out of curiosity—and sometimes to try and force pay increases at their current role—most people don't enjoy job interviews and therefore don't subject themselves to them unless they really feel that they have to. What this means is that at the time of someone handing in their notice, you're already on the back foot. Although they're still contractually present within your company, they may have been mentally present elsewhere for months.

So, what are you going to do next? First, you need to decide whether you want to fight to keep them at all. This is a function of how well they perform. Are they your top performer? Then you fight. But if you could see that your team could survive for a while without them—and that making a new hire may in fact be beneficial for the team because you're able to hire someone who is a better fit for the future—then perhaps it's better to let them go with your blessing.

The reason that you need to choose carefully who to fight for is because keeping them will inevitably introduce some form of imbalance or inconvenience for the department (or both). Typically, keeping a member of staff will involve increasing their salary. However, even if it doesn't, you may find yourself having to make other concessions that may be tricky to deal with in the short term, such as having them change team or work on projects that better suit their desires while others have to pick up the slack on the critical work. The question you have to ask yourself is this: *Are you willing to accept an imbalance to keep someone within your company?* If the answer is yes, then you fight.

It begins with a conversation. What is it that made them want to leave? You need to identify the causes. These causes could be:

- *Money.* Do they feel that they are underpaid compared to their peers or market rate? Do you think that is true?
- *Career progression.* Have they been unable to progress their career how they wanted to at the company? Is it actually possible to do so?
- *Variety.* Do they want to work on different projects, or perhaps cross-train into using a different programming language or altogether different role than they are already doing? Are the opportunities there? If not, can you create them?
- *Frustrations.* Have they suffered death by a thousand cuts where many tiny inconveniences and annoyances have become too much to bear anymore?

You need to decide whether the catalyst for them leaving could actually be satiated at your company, whether within your team or on another one. You should never fight to keep staff if you cannot actually provide the conditions under which they can become happier than they already are. You'll just defer their departure.

If it does appear that you have the opportunity to keep them, then you need to work out how to make that happen. This will typically involve calling upon others in the company around you for their support:

- *Your manager.* What do they think about the potential departure that you are dealing with? Given that they are one step removed from the situation, does your decision to fight for your staff member to stay align with what your manager thinks is best? What do they advise that you do and do you agree with them? Does the budget exist to make a counteroffer, and how does that align the staff member with their peers in the rest of the department?

- *HR.* Assuming your company is of the size to have support from HR, what do they think that you should do? If you and your manager are considering a counteroffer, how does HR feel about it? Have they got any advice on what you should do next? How does your staff member feel about the new job that they've been offered and the benefits they are going to receive compared to what is already offered at your current company? Are there additional benefits that could be extended to them so that they could stay, such as working a shorter week or working more flexible hours?
- *Other managers within the department.* Talking to your peers has two main benefits: firstly, having staff leave, no matter how experienced you are as a manager, will always suck. You can open up to your peers for their support. Additionally, if your staff member is leaving because they see the opportunity of doing more varied and different work elsewhere, are there other teams within the company that could offer them a new challenge while they continued to contribute to the company?

After embarking on this fact-finding mission, you should formulate a counteroffer. This could be a change in compensation, team, hours, type of work, and so on. Once you've done so, you should deliver the counteroffer in person to your staff member so they can think about it. You should also send a copy via email for record. Give them a few days to consider, since it may involve them talking to their family and the company that is trying to employ them.

Once they've made their decision, you'll either keep them or not. Congratulations if you did, but don't fret if you didn't. Again, you tried to do the best that you could for the company and for them, and both sides will remember that. Their new job might not be what they thought it would be, and they may be back in touch.

Making Staff Leave

Now for the harder stuff. Up until now we've spoken about employee turnover having an upside, how to let people go with your blessing, and how to fight the good fight. That's all well and good. However, did you turn to this chapter wondering how to let people go? Well, let's look at that now.

Regardless of how good a manager you are, you'll always have to deal with bad performance. In fact, it's an integral part of your job. Here's why:

- Given that you're responsible for the output of your team, if you have a bad performer, then your *output is suffering*. You'll need to work out how to deal with the situation.
- Bad performers in teams *affect their team mates*. Given that teams are interdependent, others in the team will soon become frustrated with those who are not performing at the expected standard.
- The bar that you set for acceptable performance is *firmly fixed at the level of your worst performer*. The longer you go without attempting to fix the problem, the lower your bar is as a manager. It reflects badly on you and your team.

In this section we're going to look at the Performance Improvement Plan (PIP), which is a tool that you can use to force a resolution to an ongoing performance problem. Before we go any further and look at it in detail, you should understand the following:

- A PIP should only be used when *other means have failed to improve the situation*. This means that you should not be dropping them on people as a surprise. Assuming you've been following the book and have been able to build strong and trustworthy relationships with your staff, then you should be having continual conversations with them about their performance, including during situations where it seems to be getting worse. A PIP is used when continual conversations about how to improve poor performance are getting nowhere.
- A PIP is serious, since the worst outcome for your staff member is that they are *dismissed and their contract is terminated*. Don't treat it lightly. For someone on the way out, the PIP is the last thing that happens before they are let go.
- Employment and labor laws may differ in your country, so do check the correct course of action for poor performers with your manager and HR team, or *seek external advice if you are unsure*. I have used PIPs for U.S.- and U.K.-based staff, but I cannot speak for all jurisdictions the world over.

So, what is a PIP? In short, it's a formal document outlining:

- The current situation and performance issues that are occurring.
- A number of goals that the employee will need to meet in order to pass the PIP and retain their employment and good standing at the company.

- A fixed period of time during which the PIP is active. If they complete all of the goals during that period, they have passed. If not, then the consequence is usually that they lose their job.

- How you are going to help them turn it around.

Remember that for you, as a manager, the PIP can be viewed as beneficial regardless of the outcome. If the employee passes the PIP and improves, then the performance situation has been rectified. If they do not, then they leave the company and you can then hire somebody else in their place. It's a bit like a probation period in that sense. That being said, a PIP is *not* a firing in advance. You must be committed to structuring the document and process in such a way that does actually give your employee the chance to achieve the goals that have been set out and that enables a clear understanding of why those goals are a signifier of good performance. PIPs that are unachievable are unethical and potentially unlawful. You should be as invested in your employee turning the situation around and staying as you should be in potentially finding their replacement.

Preparing the PIP

If you think that it's time to intervene on bad performance with a PIP, then do make sure that you talk to your HR team and your manager before going any further. They will have invaluable advice. Assuming that a PIP is the outcome that they suggest, then you'll need to do the following:

- *Outline the employee's current job description* for clarity.

- *Consider the performance or behavioral issue that you are seeing.* You should provide evidence that backs up your claims with reference to the job description. For example, if their work is not up to the required standard, how are you able to show that? Are they missing deadlines? Is bad code regularly being deployed to production? Have there been any formally reported behavioral incidents? Remember that you need evidenced claims, not just speculation. Is there evidence that you have spoken to them before about it? Check your one-to-one notes.

- *Consider goals that you could set for the employee to pass the PIP.* Given that you have evidence to demonstrate that their performance is not to the standard that you expect, what does the inverse look like? That is, what would prove *good* performance? Perhaps it could be particular pieces of work being completed in a set amount of time, exhibiting particular communication patterns, always turning up on time, and so on. Ensure that each goal is achievable, relevant to their performance, and can be done in a specified amount of time.

- *Consider whether training or support is required to reach the goals.* Each employee should be able to achieve the goals, so if they lack the skills to do so, then why is that and how can you support them?
- *Consider their personal situation.* Sometimes poor performance happens as a side-effect of issues in a person's personal life. This is normal, and you should be thoughtful and caring. But it cannot go on indefinitely. You can make *reasonable adjustments* to support them (for instance, changing their hours).

Make notes on all of these aspects and then share those notes with HR and your manager so that they can scrutinize them to make sure that they are clearly stated, fair, and achievable. You should ask yourself how long you think it is going to take for them to improve their performance, and how short it could be so that they don't incur fatigue. You should then discuss a reasonable timeframe for the PIP to take place in. Typically anywhere between one to three months is reasonable, similar to how the probation period works at the beginning of employment.

Creating the PIP

Assuming you've gotten agreement from others on the content, it's time to write it up. Since documentation is a key part of the process, you'll need to begin the process by formally writing it. This is most easily explained by example.

The following sidebars will contain the content of an example PIP for a hypothetical employee called Stephen. His performance has been declining over time, which has been demonstrated by his unexplained absences from work and also not completing work at the speed and standard expected from someone in a senior position.

Let's look at how the PIP could begin. It assumes that you have formally gathered feedback from the whole team about them first.

Example: PIP Introduction

This is to inform you that you are being placed on a Performance Improvement Plan (PIP). The PIP is to define our areas of concern with your performance, to reiterate our expectations, and to allow you the opportunity to improve to meet those expectations. What follows is a description of the situation.

During the last six months, your performance has become a concern to this team. Despite your seniority, you have consistently demonstrated behavior that has not been to standard, and the quality of your work has also suffered. The evidence for this is as follows:

- A recent unexplained absence from work of four days in January without notifying your manager or the team.
- Missing at least three weekly team meetings by not getting to the office on time, despite being offered the option of dialing in remotely.
- Being unresponsive to essential team communication for over twenty-four hours while you work from home.
- Being unable to meet your sprint commitments over 50% of the time during the period, often due to unexplained absences.

Your teammates have repeatedly raised this behavior as it has been affecting their ability to get their work done on time. The purpose of the PIP is to enable you to begin to perform at the level expected by the company and those in this team, ensuring that you work in a professional, transparent, and collaborative manner.

As you can see, it outlines:

- That the employee is being placed on a PIP.
- Why this is the case, with specific, evidenced examples.
- What the desired outcome of the PIP is, with the focus being on the fact that it can be used as an opportunity to improve.

Let's continue looking at the example PIP, see the sidebar on page 162. We'll continue with the objectives that Stephen has to meet to pass it.

The objectives should be measurable with evidence and achievable, assuming that the employee is serious about passing the PIP. Where possible, use exact times, days, and actions within the descriptions.

Example: PIP Objectives

These are the objectives that we expect you to achieve to complete the PIP. Each of them is coupled with a timeframe in which they should be achieved.

- Complete work that you have estimated on time, letting the team know at the earliest opportunity if that isn't possible. (three months)
- Agree on a line of communication for days in which you work remotely where you will be responsive within working hours. Be responsive via this method in times of need. (three months)
- Be on time for all weekly team meetings. (three months)
- Inform the team of any booked vacation days and ensure that work has been handed over before you leave. (three months)
- Inform your manager before 10:00 a.m. on any days that you need to take off due to sickness or other reasons. (three months)

You will be on this PIP for three months, ending on September 4. We will discuss your performance against these objectives weekly. I will also be soliciting peer feedback from the team during this time.

If you are unable to meet the objectives outlined in this plan and improvement is not possible, then we will discuss further actions, which may be, where reasonable, an extension of the PIP or possibly terminating your contract at the company.

We hope that together we can have you achieve these goals, and we are always available to help and discuss any concerns that you may have.

At the end of the document should be blank spaces for the following, which will be completed after you have a meeting about it. This space will contain a write-up of:

- Any training or additional support you agree upon through the duration of the PIP.
- Any further clarification of the objectives (if necessary). For example, there may be some additional specificity around what work being completed "on time" means, such as particular deadlines, or perhaps estimates of the rough number of days to complete a number of story points.

Additionally, you'll both need to sign it.

Delivering and Implementing the PIP

Once you've written up the PIP document, share it with HR and your manager to ensure that it meets requirements. Assuming that it does, then you need to schedule a meeting to go through it in person with the employee.

You should then do the following:

- *Set up a meeting* to discuss the fact that they are being put on a PIP and what this means. It may be the case that they haven't been on one before and aren't aware of the process, so do provide explanation and links to further documentation if you have it.
- *Share the PIP document* so they can read it beforehand and come to the meeting ready to discuss it.

In the meeting, you should calmly read through the document with them. Listen to their opinions and concerns, and if there are any parts of it which could be described better or tweaked slightly based on your conversation, then you should do so—after all, it's in both of your best interests for them to be able to succeed.

Assuming that they're in agreement with the PIP, they should sign it along with you, and then you have formally entered into the duration specified. Then, you should enter into a weekly cadence of discussing progress: a natural place for this to happen is your one-to-one meetings. Always refer back to the document with them to discuss the exact goals.

The period can go one of two ways. The ideal way is they meet all the expectations, nothing bad happens, and they pass the PIP with flying colors. However, it's possible they don't manage to meet the objectives, and you are continually making notes of times they haven't been able to meet them, such as continued lateness, unresponsiveness, and absences with no notice given.

If this is the case, it is reasonable to give a formal warning. Given one or more breaches of the PIP, write up a letter specifying the evidence for it having been broken and share it with your manager and HR for approval. Then, follow the previously described process again: book a meeting, share the letter, and sit down to discuss it. State that this is the last chance and a formal warning that they are in breach of the PIP.

After this, if they are in breach of the PIP again, then it's time to let them go. Talk to your manager, HR, and whoever else you need to make this happen. Do the right thing *for them*, empathetically. You *did* try your best, and assuming that you have been supportive, you've done all that you can do and they should move on. Your team deserves to work with a new staff member who is able to perform to the required standard, and the member of staff on the PIP deserves a clean break. You should think carefully about how you communicate the decision with your team. Ask your HR team and your manager for support if you need it.

Layoffs

A special case when it comes to letting people go, where it has neither been the decision of you or your employee, is redundancies. The unique challenge with this situation is that:

- *They haven't done anything wrong.* There haven't been any performance issues, so this is purely a business decision, and typically for bad reasons: the company isn't hitting targets and has come into financial difficulty. So even though everyone has somehow contributed to this situation in some way, it's a decision from above that is out of your control and likely isn't their fault.
- *It's going to be a surprise to them.* Because the rumor mill can be harmful, typically a round of redundancies will be carried out at an orchestrated time. They won't see it coming. This increases the potential of conflict when you deliver the news. If you are given information of upcoming redundancies in confidence so you can prepare, absolutely *do not* leak it. It never improves the situation.
- *You don't want them to go either.* When delivering the message, it's likely that you, too, will feel upset that they will be losing their job and leaving the company. You'll need to balance your emotion so that you can be clear, direct, and supportive.

If you find yourself in the situation where you have to let people go through redundancy, you will typically be given some script to follow by the company that details the reasoning plus their severance package. If you haven't been given this, then you have the right to ask for one. You shouldn't be left to make it up as a manager.

When it's time to give the message, keep calm and stick to your script and notes. Try not to react to any anger or frustration that you receive. Instead, listen and understand, but explain that you don't have the power to change the situation. This is one of the hardest things to do as a manager. You'll get through it, but it will hurt. And remember, they'll be hurting more.

Letting People Go

Letting someone go is never easy, and you should call upon the support of your manager and HR team to provide you with the appropriate administrative, collegiate, and legal support that you require.

Letting People Go

Even if you're nervous, and even if your manager offers to do it for you, you should be the one that does the firing. You owe it to your staff to be with them through the bad times and the good, and this is the very definition of a bad time.

Here are some tips for one of the most difficult meetings that you will do:

- If it helps, prepare what you want to say by writing it down first and bringing the script to the meeting.
- Speak calmly and slowly and allow plenty of silence for them to respond.
- Be prepared for them to be angry and emotional. Stay centered and listen attentively, but remember that the decision has already been made.
- Once it's over, take some time for yourself. Go and take a long walk or get a coffee down the road. Get some space between the firing and the next activity that you do.

Enough Goodbyes Already!

There we are: everything you wanted to know about people leaving the company. It's going to happen. In this chapter you've learned the following:

- *That people will always leave.* You should now know that it's perfectly normal and there's often little you can do about it. Natural turnover, however, can be advantageous in bringing new, diverse ideas to your team.
- *How to know when people should go with your blessing.* You learned about the good reasons that people may leave for and how to accept it, to be grateful that you got to work with them, and to send them on their way with a smile.
- *How to fight for staff to stay.* You considered when it's appropriate to try your best to keep staff from leaving, what mechanisms you can use to do so, and the trade-offs you'll face.

- *How to make people leave.* You then learned about Performance Improvement Plans: when to use them, how to write them, and what to do during the process. Remember that despite the difficulty, they're a win-win situation for you—either a poor performer improves or they leave and you get to replace them with someone else.

Given that we've ruminated on all of these goodbyes, let's turn our attention in the next chapter to something more positive: how you can help people across the *whole* department rather than just your team. Doing so improves your skills, increases your visibility and influence, and lets you positively affect others outside of your team. We'll look at a number of strategies for doing so next.

In charity there is no excess.

➤ Francis Bacon

CHAPTER 9

How to Win Friends and Influence People

You wait as the coffee machine slowly fills up your cup.

"Everything going OK?"

It's Mark. He's a fellow engineering manager running the team that's responsible for dashboarding in the application.

"Yeah, all good. Definitely need this coffee though."

"Tell me about it!" Mark replies.

A voice calls from across the kitchen.

"Oh, hi Mark! Thanks so much for our chat yesterday, it really helped."

It's Abigail, who runs the European sales team. You've seen her around but have always been intimidated by her seniority.

"Oh, no problem, it was good to find out how we're doing in pitches. I've let the team know what you said."

"Awesome, catch you later!" says Abigail, hastily departing.

You're curious as to how Mark managed to make friends across the engineering and sales divide. You're just about to ask him, but another voice erupts from behind you.

"What's up Mark?" It's Cara, a back-end engineer. "Thanks for helping me fix that bug yesterday. It always turns out to be such a stupid error, doesn't it?"

"Hah, don't worry about it. You just fixed it yourself while I sat there. You should see some of *my* bugs sometime..."

"Catch you later!" says Cara, walking off toward the meeting rooms.

Mark's coffee finishes decanting, and you begin walking together back toward your desks. You remember that you were going to ask him about how he got to know Abigail, but, once again, another voice interrupts.

"Ah, just the person I was looking for!"

This time it's the CEO.

"Mark, we've got one of our biggest customers coming to visit us next week, and they'd love to meet some folks from Engineering who build their favorite features. Would you be up for meeting them?"

"Of course!"

"Great. There's an invite coming to your calendar."

She walks off. You're flabbergasted. How is he so well connected to all of these people? What is he doing that you're not doing?

You turn toward him and open your mouth to formulate the question, but there's another voice emanating from behind you.

"Oh, Mark!"

What a popular guy...

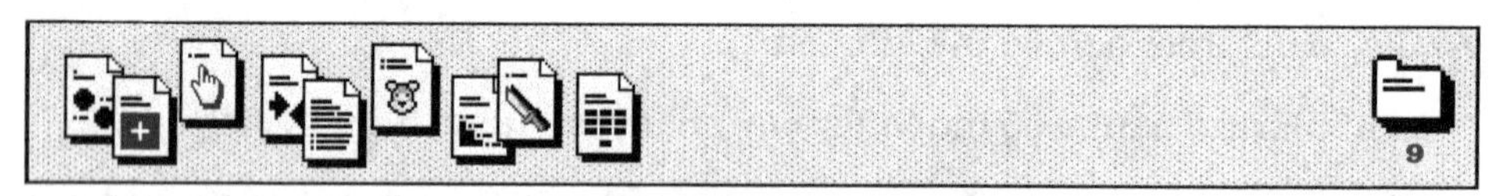

In this chapter we're going to look beyond your team: we're going outward into the rest of the department and company. Excellent managers are well connected, and being well connected carries a myriad of benefits. We'll look into these in more detail as you learn the following:

- *Why considering the world beyond your team matters.* By being more engaged in the wider department and company, you'll be able to discover more ways to get involved and that, often, many of these ways can positively impact your team. We'll give you the motivation you need.
- *How to build your network inside the company.* We'll look at concrete strategies for building your network and understand why it's important.
- *How to give back to others through mentoring and coaching.* Another way to build your network is through *giving back*. By reading this book you've already learned many valuable skills that can help others out. We'll look specifically at *mentoring* and *coaching* as ways that you can contribute.

You'll begin to see that there are so many more ways in which you make your company a better place.

Going Beyond Your Team

So do you remember our favorite equation? Surely you must know it by heart by now. OK then, one more time. Your output is measured by:

The output of your team + The output of others that you influence

This is the part of the book where we begin to consider the second part of the equation—the output of others that you influence. But what does that mean? And isn't influence an ambiguous word, or one that is potentially loaded with politics and dubious intent? You should understand that it really isn't. Influence can be an incredibly positive thing.

To get some context, cast your mind back to what we discussed in *Manage Yourself First* around the four key activities you're involved in as a manager:

- Information gathering
- Decision-making
- Nudging
- Being a role model

The third and fourth activities, *nudging* other people's decisions by offering your opinion, and *being a role model*, where you publicly act with the values and behaviors that you wish to see in others, can be seen as a couple of ways of exerting your influence in a positive way:

- You may have some past experience that allows you to nudge a decision that another team is making in a way that makes them get it done more effectively. That's *nudging*.
- You may continually give feedback in an open and candid manner that encourages others to do so within your team. That's *being a role model*.

So given that you are additionally measuring your output as a manager as the output of others that you influence, it follows that by having more opportunities to influence others, you have more opportunities to increase your own output. That's why being well connected is important. It simply means opportunities to have a positive impact on more people.

The curious corollary is that influencing others positively builds stronger connections. This allows you to gather more information about what is going on in other parts of the department and the business, feeding that back into making better and more impactful decisions within your own team. And guess what? Those are the other two managerial activities. It's a virtuous cycle.

But if you're new to a company, or even if you've been promoted from within as an individual contributor, then you might not be particularly well connected beyond the individuals within your team and your manager. You can't expect your manager to make all of the connections for you either. It's going to take some concerted effort. It can also be difficult for those that are natural introverts (guess what: that's me); but being well connected strengthens your position as a manager, positively impacts your team, and most importantly of all, can help you make new friends and feel supported at work.

Let's start by actively building your network.

Building Your Network

We work in organizations of varying sizes, from the very small (for example, startups) through to multinational corporations with tens of thousands of employees. As a manager in the engine room, you have some idea of the general pulse of the business, as you are doing work that serves some business need, whether that's building new features, improving the infrastructure, or fixing bugs. You may also involve staff from other areas of the business, such as stakeholders in your projects, and have a chance to engage with them there. Yet, this doesn't really give you the whole picture. The context of your work and conversations is likely bound by the project that you're currently working on and the specific stakeholders who are keen on it getting built.

It's advantageous to begin to experience a more rounded picture of how the business is doing. In *Only the Paranoid Survive [Gro98]*, there's a concept that the *snow melts at the periphery*. But what does this mean? It means that the employees on the periphery of the business—for example, the salespeople going up against competing salespeople, and the customer success staff working directly with your clients—see the *real* situation in the market with your products much quicker than you do; you're simply further away from the action. This is especially true at large corporations. It follows that having good lines of communication with key periphery staff can make a business react quicker to market conditions. The larger the business, the larger the void between the periphery and the creative core. This means that smaller competitors have a speed advantage by default. If you're interested in models of disruption that allow startups to win, then *The Innovator's Dilemma [Chr16]* is a fantastic resource.

However, this isn't just about startups biting at your ankles. This is about being a better manager. You can build relationships with your peers in Engineering and also with those who stand where the snow melts—which is beneficial for both you and your team. If you were to imagine a situation where you have regular contact with others in the business, then you have many more opportunities to help people and thus increase your own output:

- You can experience their challenges with customers and competitors firsthand and speed up or lobby improvements to your software, especially if you identify that there are quick wins (for example, features that require little engineering effort for a large amount of customer impact).
- You can build rapport with staff in other departments who can become influential stakeholders in your future projects.
- You can be party to strategic discussions as a stakeholder from Engineering, which again gives you a chance to offer your opinion and nudge decisions.
- You can increase your profile in the company, which gives you a better chance of being considered for new opportunities that can grow your career.
- And guess what? You can make some new friends. Who'd have thought it?

Your Turn: Who Could Be in Your Network?

Take a moment to think about other people that work in your company who you think would be beneficial to have more regular contact with as part of your network. Who are they, and why would they be a useful contact?

If you are finding it hard to come up with a list, then perhaps these questions might help guide you. Think of...

- Someone in your department that you are impressed by or look up to.
- The most visible or successful salesperson in your company.
- Somebody who is involved in marketing your software.
- Somebody in your company who is a power user of your software.

Who comes to mind? Perhaps they should be first on your list.

Making Introductions

There's no standard process for being introduced to the right people within the business. You'll need to get creative here. If you're new to a company and don't have a clear network, then you can ask your peers or your line manager who they typically interact with and why.

- When conversing with your *peers*, you can ask for a list of staff in other areas of the business that they have had a positive experience with, either informally or through being stakeholders on their own projects. Also ask them who they think are the main decision-makers in other departments and who they see as particularly influential and interesting.
- When conversing with your *manager*, ask who their own personal network is outside the department and whether there are people within that network that you should be introduced to. Try and find a diverse set of individuals with differing interests. For example, can your network spread over multiple departments, geographic locations, genders, and seniorities? Having a diverse network lets your ear hear many varying opinions, which can make your own more balanced and better informed.

If you haven't met the people that were recommended to you, then you can always introduce yourself via email. Don't worry about emailing out of the blue: when was the last time a colleague sent you a polite message introducing themselves and it caused you to have a negative reaction? Hopefully never—although I'm sure it's an interesting story if it did happen.

Let's have a look at an example email that you can send to the list of people you created, see the sidebar on page 173.

You can replace the questions with anything that you feel is a current hot topic that is of interest to you, so take those that are present as examples only. Some other questions that you could ask could be:

- What they think of the most recent project that your team worked on.
- What their favorite feature or recent improvement is.
- What their dream feature would be (although, do make it clear you're not just going to go off and build it...).
- What their favorite software is and why.
- Who they look up to, or have learned a lot from, at work.

The list of questions is only limited by your imagination. Go wild. It's scary and fun getting to know new people.

Example: Introduction Email

Hello,

Apologies for the email out of the blue. I'm X and I manage the Y engineering team. We're working on Z.

I'm beginning a practice where I'll be frequently checking in with others within the business, with the aim to extend another hand outward from Engineering. If you're receiving this, I'd love to know whether there is anything that I can help with or discuss!

Here's a starter for ten to get the conversation flowing:

- Do you feel that there is anything we can do to improve our communication outside of the department, notably on what we're planning, working on, and shipping?
- Is there anything pressing that you would like more information on or any issues that you'd like to raise?
- What could we do better as a whole?
- Additionally, are there any projects or discussions that could use my input?

If you don't reply, then that's not a problem—I'll just assume that everything's going well right now.

Best,

The Intelligent Reader of This Book.

Checking In Regularly

Once you have a list of folks that you've created an initial connection with, it's up to you to decide how you would like to interact with them. You have options for the regularity and formality of your check-ins.

A formal way is to get in contact via email once a quarter, then follow up with all of the points that are raised. It helps to offer a leading question to get the conversation going, as demonstrated earlier. There may be a new product launch to discuss or an announcement of the upcoming roadmap. It's easy to capture and distribute the information when using a medium such as an email.

However, it's also extremely valuable to hear the softer side when interacting with your contacts. Informal meetings with those that are geographically closer, such as grabbing a coffee every month, can be a great way to inform each other, vent a bit, and learn something new. They don't need to have an agenda. Trust the process and see what emerges. You can do virtual coffees over video chat if you're in different locations. This information can feed into your own decisions and make them more balanced, informed, and considerate. If your company has different offices, if you ever need to travel to them, it's nice to potentially have a friendly connection there before you arrive.

As you get to know your network more, you'll be able to continually expand it. Conversations will naturally mention other interesting people whom you can get in contact with, and the process will repeat ad infinitum. Soon you'll be surprised just how many people you'll be beginning to get to know.

Give it a go: get in contact with those outside your immediate circle and see what happens. I think you'll be surprised. After all, people love to talk.

Giving Back

In addition to making connections with others around the company to learn and to share information, you've already learned a multitude of skills in your career and by reading this book that can allow you to make additional connections by sharing your time and skills.

You can do this in two ways:

- *Mentoring*, which is where you can provide specific advice and teach certain skills to somebody.
- *Coaching*, where you help others explore problems and come to their own solutions.

You may hear both of these terms used interchangeably, but they actually are different. The relationship between a mentor and mentee is more similar to the relationship between a teacher and a student. Most people have experienced a form of education, so it's easy to visualize how that relationship could work. However, not many people have been coached, so not many people know how to do it, nor do they even know when they are being coached!

We'll explore both mentoring and coaching in more detail so you can begin doing it yourself and, consequently, encourage a culture of it in your workplace. It's extremely rewarding to be able to give back to others, especially if they're in other parts of the department or company.

Mentoring

Mentoring is a relationship in which one person shares their experience, skills, and knowledge with another person in order for them to progress in their career. Mentoring relationships typically last for months or years, with the mentor becoming an accessible point of contact for the mentee. Sometimes a mentorship can last for a whole career.

We call mentoring a *relationship* because it is more than just being a designated person to offer occasional advice and feedback. Instead, a mentorship represents an investment in the career development of the mentee from the mentor. This means that the relationship has structure: regular meetings, an environment of encouragement and empowerment, and a focus on achieving goals.

You have a number of skills that you could already be teaching via mentorship:

- If you were previously an individual contributor, then you have a number of skills that you could teach to a junior engineer.
- By reading this book you've already learned many skills that could allow you to be a mentor to another brand-new manager.
- You may have desirable supporting skills, such as being a seasoned speaker, writer, or open source contributor.

You may be thinking that there are many parallels with being a mentor and being a manager. This is true. This is also why you're in a good position to do it. It also improves your own skills, so there are plenty of reasons why it's beneficial to be a mentor:

- It allows you to develop your skills as a manager, even if you're not managing the individual.
- You get to increase your impact within your company as a positive contributor willing to help others, even if they are on other teams.
- You'll learn about the kinds of challenges that people are facing in other teams.
- You'll get to know other people that might even want to come and work on your team in the future.
- Most importantly, it's extremely rewarding, and it will make you feel like you're making a real difference.

For the mentee, there are plenty of benefits also:

- They can learn new skills and knowledge.
- They too can widen their network with the company by having a close relationship with somebody outside of their team.
- They can use their mentor as an impartial third party: they're not talking to somebody who is accountable for their performance or their pay.
- They can use their mentor to gain distance from their day-to-day work and approach problems more abstractly.

Finding People to Mentor

Since a mentorship relationship should be initiated by the mentee, you shouldn't just go around thrusting your desire to mentor on others. Instead, you need to work to create an environment in which these connections will happen naturally. If there are mentoring relationships already happening, then it may already be standard practice for managers to recommend a mentor for their staff, or for those seeking a mentor to reach out to matchmake with others.

However, if there is little mentorship happening at your company—and a quick straw poll between your colleagues could confirm this—then there's a way in which you can get it started.

Creating a Mentorship Agreement

Assuming that connections get made within your department—and I'm sure that they will—then it's important that both the mentor and the mentee enter into the relationship by clearly establishing what they would like to get out of it up front. You can do this in an almost identical way to how we approached the contracting exercise in *One-to-Ones*.

Below are *five questions [Cam19]* that a mentor and mentee should consider before having their first meeting. You can do so in a shared document so that you can collaborate.

- *Time*. When is the relationship starting and ending, or is it indefinite? How often will you be meeting and where? Is this something formal in a private meeting room, or is it a conversation over a coffee or during lunch?
- *Flexibility*. Is it acceptable for the mentee to contact the mentor at any time via any means (for example, via in-person interruptions or video calls) or should communication be batched in a weekly meeting?

Your Turn: Creating a Mentorship Matrix

You have the power to kick-start a culture of mentorship in your department, and all it needs is a line of input from each person.

What you can do is the following: create a shared spreadsheet with the following headings, fill it in yourself, and then share it with your colleagues, or even the whole department. If doing the latter, see if you can get some buy-in from your manager to promote it more widely.

Each row should have the following headers:

- The person's name
- Their job title
- Their team
- Their contact information (email, chat handle, and so on)
- What skills they are seeking in a mentor (for example, management skills, public speaking, JavaScript, and so forth)
- What skills they are able to offer as a mentor themselves

Once you're done, send it around with an explainer that highlights that participation is optional, but there are a multitude of benefits to being a mentor and mentee, as you've already learned in this section of the book.

Give it a go: you might find yourself with a mentor and mentee before you know it!

- *Confidentiality*. You should both agree that the content of the meetings is confidential unless specified otherwise. If there are particular elements that aren't confidential, then do say.
- *Boundaries*. Given that it is unlikely for the mentor to be managing the mentee, it's important to be clear that there are some boundaries with the relationship. For example, if the mentee is under pressure to complete some work to a deadline, then it's not the mentor's responsibility for it to get done. Instead, outline what the relationship offers (for instance, learning and discussing particular skills) and what is outside of those boundaries.
- *Review and evaluation*. If you're entering the relationship for a fixed period of time, then discuss how you will review the relationship at the end and decide whether it should continue or terminate. If the relationship is indefinite, then it's useful to schedule in a checkpoint at which to review it regardless, so that you can give each other feedback.

- *What if it doesn't work out?* Decide on the way in which you will inform each other if the mentorship isn't going to plan, and what signs that you will look for in each other if that is the case. Additionally, agree on whether there are particular issues where the mentor should have permission to get in touch with the mentee's manager for escalation.

By having the mentor and mentee consider and answer these questions ahead of the first meeting, it will ensure that the mentorship gets off on the right foot, defined by the boundaries set by both of them. It facilitates discussion that makes for a more successful and transparent relationship.

Your Turn: Get a Mentor and Mentee!

See whether you can find a mentee over the coming months. You may do so via sending round a mentorship matrix that advertises your skills, or perhaps you may offer your support to someone.

You should also see whether you can get a mentor for yourself, potentially to further improve your management skills. Is there anyone in the department that you look up to as a skilled or senior manager. Do you reckon that you could ask them to be your mentor?

Coaching

So, you've learned about mentorship: it's a structured relationship between two people, one with skills and knowledge to teach, and one that would like to learn them. However, you can still form relationships with people and greatly assist them in their problems, even if you don't share any skills with them. You may find yourself wondering how you can have a tangible impact on the performance of people of *all* skill sets and *all* experience levels.

- You may already have people on your team that you don't share a skill-set background with. Up until now, you've been able to focus your conversation within the bounds of what the team is working on and trying to achieve, and support them through it, but how do you really help them solve any problem?

- You may have staff on your team that have many more years of experience and expertise than you do. As such, you may have been struggling to give them direct advice, as they're able to work things out on their own most of the time. How can you still help them develop?

You might hypothesize that if you're able to learn a methodology to support those that have different skill sets and even more experience than you, then you can even give that support to pretty much *anybody*, even if you know nothing about how to do their job. And you'd be right. That's what *coaching* is. It will allow you to offer support to anyone within your team, and if you get good at it, you can give back to the rest of the business by offering coaching to colleagues whether they are in sales, marketing, product, UX design, or engineering. Sounds good, no?

So what is coaching? At an abstract level, it's a technique for helping your staff improve their performance. But that sentence is so lofty it almost means nothing. My own take on coaching is that it's a *framework for interactions* with your colleagues that makes them more likely to have tangible positive effects in their work.

Before we dive in, let's make this clear: coaching doesn't require special planning, skills, acting skills, or, well, anything really. You don't need a whistle and a tracksuit. It's simply a collection of tools that you can use when having a conversation with someone. You can use these tools in formal settings such as your one-to-ones with your staff, or you could use them while having a chat about their work in the kitchen. Equally, they can be used while in conversation with one person, or they can be used to steer a group discussion.

If you try them out, you'll be surprised at how useful they are. Remember this: essentially you can coach someone at *any* time by having a particular type of conversation. Whether you enter a formal coaching relationship or not is up to you, but if you master the coaching *mindset* as a manager, then you can have a positive effect on others with almost every interaction.

Modes of Conversation

So if coaching is effectively just having a structured conversation, is there a special way in which you should be talking to the person being coached? The answer is yes, there is. However, it's straightforward to grasp.

To frame it, consider *two mutually exclusive modes [Dow14]* of conversation:

- *Directive*: This is where you're instructing someone on what they should do. This is considered to be a "push" action—that is, you're solving their problem for them.
- *Following interest*: Here you're predominantly listening to understand, reflecting on what they are saying, and summarizing. This is a "pull" action—that is, you're helping them solve their own problems.

When having a coaching conversation, you'll need to keep these two modes in mind and use them to steer the conversation in the right direction.

If you think about mentoring, you can see that it's a relationship that is primarily *directive*: the mentor is teaching skills to the mentee. However, with coaching, the primary focus is trying to shift the conversation to *follow their interests*, which magically allows them to solve problems all by themselves.

In reality, a coaching conversation will swing between these two different modes. You'll purposefully use sentences that are questions, suggestions, and summaries in order to follow their interests and encourage them to come to the conclusion on their own:

- "What's the difficulty with doing it in that way?"
- "Why did you pick that method?"
- "What's up with Bill this week?"
- "Why's that?"
- "Tell me more about that API."
- "Walk me through your thought process."
- "Can you draw me a diagram on the whiteboard?"
- "So what you're saying is..."
- "So am I right in thinking that..."

As a coaching conversation progresses, you'll work out whether you need to be directive to unblock some thinking, or continue to follow their interest until they get there themselves. You may find that less experienced staff require you to be more directive than their more experienced colleagues. This is normal. With time, as your staff become more experienced, you'll find yourself following their interests much more than you'll find yourself directing them. The positive side effect of this is that as you become more experienced yourself as a coach, you realize that you can help anyone out with pretty much any problem, as you'll be mostly listening and asking questions!

The rule of thumb is this: at each point of the conversation, be conscious as to whether they are honing in on the answer and keep listening and suggesting until they do. If you think they'll never get there, be more directive.

Keep the Thought Bubble over Their Head

Unless being explicitly directive, to help your staff develop their skills, both in technical and abstract problem solving, an excellent tool for following their interests is *keeping the thought bubble over their head.*

Here's a little example.

You: "So how's progress been this week?"

Them: "Not great. The back end is much harder than we expected, and we're not sure how to approach it."

Even if you're the world's most expert engineer, don't tell them what to do if you want to practice coaching, especially if you know what the solution is. Instead, every time you feel that you would naturally jump in and solve a problem, imagine yourself pushing a big cartoon thought bubble away from your own head so that it sits over theirs.

You: "OK, so what's hard about it?"

Them: "We were going to store it in MySQL, but there's way too much data and it'll be slow when we access it, as we'll have to scan every row each time."

Now, at this point, you may know exactly how to solve this problem. But you shouldn't. Push the thought bubble back to them.

You: "Interesting. Have we solved any storage problems like this before?"

Them: "Hmm. When Alice's team were storing alerts a few years ago, I remember they couldn't use a relational database either."

You: "That project was quite successful. Do you know what they used?"

Them: "I don't. But I'm going to go and grab her for a chat after this meeting and find out."

Note that all you did was ask some fairly open, leading questions. Not only did this allow them to arrive at the next steps on their own, it coaches an approach to thinking about problems that your staff can reuse in the future. Additionally, it encourages thinking through problems in the open, together, collaboratively.

Now, if they were totally stuck, you could be more directive, but in a way that still lets them figure it out on their own. Let's replay that last exchange:

You: "Interesting. Have we solved any storage problems like this before?"

Them: "I don't know. I'm stuck."

You: "I've got a feeling that Alice's team did something similar. I think you should have a chat with her."

Them: "OK. I'll try and find her after this."

Your Turn: Be a Coach in Disguise

This week, apply the coaching principles that you've learned here to your one-to-ones. Instead of trying to offer solutions to problems, instead push the thought bubble over their head and turn directive conversations into ones that follow their interests so that they solve them on their own.

After giving it a shot, what did you notice about the types of conversation that you were having? Did you feel that you were able to be more generally useful in your support for your staff? Did any of them have a eureka moment entirely by themselves?

Forming Coaching Relationships

In the same way that you can form a mentorship with another member of staff, it's also possible to form coaching relationships. Once you begin to get comfortable with performing coaching conversations using the techniques that we've covered here, then you can offer coaching support to anyone, regardless of their skill set.

Coaching relationships can be long-lived, or they can even just be a one-off session to tackle a particular problem. For structured coaching sessions it can be useful to frame the exchange in a sequence of stages to ensure that it stays on course. A good approach is the *GROW model [Dow14]*. For a given topic or problem, you'll want to iterate through:

- *Goal*: What's the goal of this session? What problem are we trying to solve?
- *Reality*: What's the situation like now? Who, what, where, and how much?
- *Options*: What are all the different ways in which we can tackle this issue?
- *Wrap-up*: Become clear on a choice, commit to it, and discuss what support is needed.

Entering a coaching session with the GROW model in mind will help you keep the conversation on track if it begins to deviate. Let's think about the previous example using this model.

- *Goal*: We need to work out how to store the data in some way that's fast enough to support our use case.
- *Reality*: We know a lot about relational databases, but they won't be fast enough for this problem. We have a hard deadline and we need to get some support.

- *Options*: We could get an external consultant, but that's very expensive and they might not be available immediately. We could try out some different databases ourselves, but we're not expert enough to make an informed decision about what to use in production. We could see whether anyone else in the company has had to solve a similar problem and seek their advice.
- *Wrap-up*: We've decided to seek advice from Alice's team because they may have implemented a similar solution in the past. We're meeting with her later, and we'll loop back with a decision on how we're going to proceed. Then we'll make a call on what to do.

Your Turn: Run a Coaching Session!

Look for an opportunity where a member of staff is stuck with a problem, and offer your support coaching them through it. Tell them you're practicing your coaching skills.

Book some time in the calendar and run the GROW model through with them, perhaps by writing it on the whiteboard. Then, have a coaching conversation with them. Once you're done, reflect on how it went and how you can improve it next time. Did they make progress on solving the problem by talking to you? Did they have clear next steps after you had your discussion?

Coaching Anyone in the Company

Directing your conversations using a coaching framework will greatly improve the quality of your interactions with your own staff. You'll notice that with time you'll get better at doing it, and similar to driving a car or riding a bike, you'll just do it naturally.

If you get really good at it, you'll notice that you can pretty much talk about any problem with anyone and assist them in thinking it through. This can open up an opportunity for you to increase your influence in your department and company: you can coach others that don't report to you. If you've demonstrated to your peers and your manager that you are a good coach, why not ask if there's anyone that they would recommend starting a regular coaching session with you?

Being a neutral third party in the coaching relationship, rather than someone's line manager, means that you can form a close connection with the person that you coach without the worry that you ultimately need to judge their performance. It also allows you to have a positive impact on other parts of the business without feeling that you're meddling in their affairs. Instead, you're enabling others to solve problems for themselves by working them through with you. It's very rewarding.

Receiving Coaching

As your career progresses, you may find that you need more coaching support than your company can provide. Your line manager may already be very busy, frequently traveling or based in a different country. Additionally, if you become a senior manager, such as a VP or a CTO, you may be wanting help working through problems that could be too sensitive to share with your peers or your direct reports.

In that case, it may be beneficial to ask whether the company could set you up with an external coach, who you can meet with regularly to unpick the knots around issues in your brain using the same techniques that we've described in this book. Prices vary and can potentially be expensive, but there's the possibility that the executive team may have connections that could offer some coaching services for free or at heavily discounted prices. Just ask.

Time to Take It Up a Level

And there we are. You've learned a great deal in this chapter. We've covered:

- *Why considering the world beyond your team matters.* We revisited our favorite output equation and saw how building your network and assisting others outside your team can positively affect your output as a manager.
- *How to build your network inside the company.* You learned about how the snow melts at the periphery, and how to build your network so that your metaphorical eyes and ears can see and hear better to help yourself and your team.
- *How to give back to others through mentoring and coaching.* Finally, you learned about mentoring and coaching, and have the tools to mentor and coach others and find mentorship and coaching for yourself.

By practicing these skills, you'll improve your ability to help others, become better connected, feel satisfaction at your ability to give back, and you might even get interrupted as much as Mark.

This also signifies the end of the second part of the book. Congratulations for making it this far. We're going to move into Part III, which is where we begin to tackle some challenging and meaty problems. We're going to start by looking at the stress and pressure that you and your team will feel and ways in which you can cope when your back is against the wall.

Part III

The Bigger Picture

In addition to building your relationships with individuals, you'll need to widen your view to how you operate in the workplace as a whole.

We're going to look at the bigger picture. We will explore the difficulties of working with other people and on important projects. We'll spend some time considering office politics and whether you should take part. Then we'll be introspective and consider stress and anxiety.

Diversity, flexible and distributed working, and culture feature later. Lastly, we'll learn how to construct career ladders and then consider your future and where it may take you.

Those people who think they know everything are a great annoyance to those of us who do.

» *Isaac Asimov*

CHAPTER 10 Humans Are Hard

Your team is getting it in the neck today. After months of technical exploration and prototypes, they've been unable to produce anything that looks like it will make the back end of the new product work in the way that it was imagined. The team is frustrated and is convinced that without spending potentially millions of dollars in hardware and annotator time, they're never going to be able to deliver anything of value.

This isn't sitting well with the VP engineering and the sales director. In fact, they're completely livid.

The sales director is first to interject. "You're honestly telling us that after six months you have nothing that is going to get this thing built?" Your engineers aren't used to this kind of intense confrontation, especially from such senior people. It makes them feel like they're pretty stupid.

"We tried. It's just a really difficult problem."

Their nervousness makes them fail to say they've produced some interesting prototypes and insights into tangential problems along the way.

"Six months and six people and we have nothing? That's the problem with this team. I just don't think you're working hard enough, especially when compared to the sales team. They often work from 7:00 a.m. to 7:00 p.m. When are you all going to step up and take ownership?"

"I'm sorry. It just doesn't seem possible," replies your data scientist sheepishly, again failing to mention that they've produced a number of useful prototypes that could eventually make their way into the product.

Your VP engineering is gripping her pen tighter, as one of her teams is making her look stupid in front of her peers. "You do realize that other companies are just solving this with machine learning now? It's all off the shelf. Should take a couple of weeks tops. I've seen our competitors launch stuff really quickly this way. Why can't you?"

"It's not that straightforward; it's really hard to get right."

"Others are doing it. Did you not see Compusoft's press release last week? I'll be blunt: I just don't think you're working hard enough. I reckon we should have just outsourced this problem. That would have got it done in half the time."

Everyone is angry and at loggerheads. Senior management are disappointed with performance, and the team on the receiving end feels dumb and unable to demonstrate they've done worthwhile work.

How did you get here? Why are people acting in this way? What should you do as a result?

As you may have begun to discover, humans can be challenging. More challenging than software. Even more challenging than a computer science degree. Not only are the humans on your team tricky, the humans that are outside of your team can be even trickier. This chapter explores the difficulty of working with humans and digs into why situations like the one you just read occur.

Here's what we're going to look at:

- *Scrutiny and judgment*: Your increased visibility as a manager invites increasing scrutiny from others. We'll look at what it means to be a leader and what it exposes you to, and how you may find yourself scrutinizing others.
- *Wobble*: Your company is like a Jell-O. If everything goes wrong at the top, it wobbles all of the way down. We'll look at this concept and how to contain it.
- *The whip and the carrot*: Those outside of your team and department might want you to make your team go faster in the same way jockeys whip a horse in the final furlong. Instead, you'll want to let them lead themselves toward the carrot in a way that promotes their growth and happiness.
- *Mount Stupid*: Regardless of their true intent, humans can do and say things that are unintended, misinformed, or sometimes just plain daft. We'll look at common misconceptions and the psychology behind them.

Let's kick things off with the perils of the limelight.

Scrutiny and Judgment

These days, you wonder why anybody would want to live their lives in the spotlight. Politicians, public figures, and businesspeople are deeply scrutinized by the press and members of the public.

The media publishes a vast array of content about the minutiae of these people's lives: where they are currently on holiday, and of course, not forgetting the conversation about whether they should be on holiday at all. There's snide commentary on a mistake made during a public appearance. And how dare they wear that particular outfit?

These people are only human. However, their visibility means that they are subject to all manner of nitpicking in the space between what they are perceived to represent and the reality of their lives. It must be utterly exhausting. But how many of these scrutinies are unique to public figures?

Perhaps we could take the examples in the previous paragraph and reimagine them in the context of the workplace. Why is it that the CEO is currently on a hiking vacation with her family when the whole company knows that they didn't hit their target this quarter and there are serious issues in the sales organization? How inept can our VP product be if he can't even get his facts right during the presentation of the roadmap at the company meeting? How can it be possible that our CFO won't let any of her team work from home more than one day a week when she works from home two days a week herself? The scrutiny of those in positions of power can be just as prevalent in the office. Fingers will always point.

As you spend more time as a manager, and especially if your career takes you further down the management track (see *Dual Ladders*), you'll find that increasingly senior positions invite more scrutiny from your colleagues. Going up the org chart from an engineer to a manager, to a director, then VP, SVP, or C-level position has a similar effect to those who take up seats in public office. Instead of just being a contributor to a team, you begin to represent the team conceptually as their leader.

At more senior levels, whether you like it or not, your role can expect you to embody particular values that the company holds dear, as you act as a role model for the whole department or organization. During good times this can be fantastic: the company is thriving and you're the living embodiment of all that is wonderful and successful. During bad times, you'll get the fingers pointed at you as you are fundamentally accountable, even though it may have not been your fault.

We all want role models. We want people to look up to who we can unite behind and put our faith in. These people set an example for how the rest of us should act. Think of the involvement of public figures during times of national grief or crisis. We have an expectation that they will be present at particular events, that they will say the right words, and that the correct symbolic gesture will be performed: the laying of a wreath, the cutting of a ribbon, the giving of a speech. This accountability naturally breeds scrutiny when the people in the position of role models inevitably demonstrate that they, too, are only human. They make mistakes, they get angry, or they do something stupid, just like everyone else. Like those public figures, we can't expect perfection from those above us in the organization. It will only lead to disappointment.

As Above, So Below

You will encounter scrutiny and judgment from all directions: from being on the receiving end of scrutiny from those that report to you, and also feeling those same feelings *yourself* toward those you report in to (and beyond). Where does it come from and how do we deal with it?

Let's look at both of these situations in turn.

When Your Team Scrutinizes You

To err is human. You, too, will have bad days as a manager. You'll have days where you will lose your sanity and get angry or stupid or both. You may do something that conflicts with a principle that you've preached before. Your team may totally mess up a product launch. It happens. However, these conflicts between your conceptual position in the organization and being a fallible human invite a variety of emotions from those that report to you.

- *Kindness and understanding*: This is the best possible reaction to come from your staff. If you're having a bad day or a bad week, they understand and they want to be on your side making things better. They may ask if there is anything you can delegate, or they may just take you out for a coffee and a chat to see if you're OK. If this is the response that you get, then know it's due to two reasons: you have fantastic staff and you're *doing a good job.*

- *Concern and worry*: Those that look up to you for stability will feel your situation deeply and potentially catastrophize over the real meaning of it, even if that meaning is something entirely of their own invention. You might have had a terrible night's sleep and are a bit grouchy, but they interpret it as you disliking them and their job is at risk. If this is happening, you should lend your ear and your time. Be candid about what happened or why you're feeling this way. Reassure them it's nothing to do with them. If there's a way for them to practically make the situation better, such as through delegation of work, then they can contribute to making things better themselves.

- *Resentment*: They begin to turn away from you emotionally. "Why is it that I have to work for a leader who is not representing me better in the organization? How could they make me do that? How can they be paid more than me? I do all of the work!" Understand their anger. Ask them why they feel the way they do. Dig deep in your conversations. There will be a reason, and it often isn't entirely you. It could be related to pay, career growth, daily frustrations, and so on. The list is unbounded. You can open that conversation about it.

- *Mutiny*: They fully turn and can begin to sabotage you. They're not on your side both privately and publicly, and you feel that they're trying to throw you under the bus. Maybe they feel misrepresented and are rebelling, or maybe they feel they could do your job much better. This is the *danger zone*.

The further down this list that your staff may be, the more work it's going to be to get them back on your side. As with most issues with interpersonal relationships, open and honest communication is the remedy. Stressful times make suppressed emotions come out of the woodwork. It's your job to accept that this will happen and talk through them.

Your Turn: Embracing Scrutiny from Your Team

Next time that something bad happens, or if you're going through a stressful time with your team, make a conscious effort to note how each of your staff feels toward you. Where on the emotion scale are they? Do they rally around you to help, or resent it?

For those that are nearer to the bottom of the scale, do the following:

- Take the extra time with your staff to listen to how they feel. Extend your one-to-one or go for a coffee. Try to understand how the situation is affecting them as a person.

- Remember that often you are not the cause, just a catalyst. Try to find the root cause of their frustration. What is it?

- Try to frame the situation positively: you're getting a chance to talk about a gnarly issue!

What did you find out? What was the real cause of their feelings toward you?

When You Scrutinize Your Manager

Even if you've mastered the art of identifying and remedying how your direct reports react to your own wobbles, it's natural to feel any or all of the same emotions toward those who are above your own position in the org chart. You work with them closely, you know their strengths and their weaknesses. You've seen them on good days and bad days. And get this: if you report in to the executive of a public company, you may feel even more resentment if they're having a tough time; you have access to their salary data!

As hard as it can be, you will need to align your misgivings with your own manager—assuming that they are not due to serious misconduct—from a place of empathy. They are subject to even more unstructured debate and uncontrolled emotion than you are. They may have family and dependents that pull them in multiple directions. Approach them with kindness and understanding and ask whether there's anything that you can do to help them. There may even be the possibility for career progression if they begin to delegate more to you through your helpfulness. Don't let resentment turn to mutiny: it never ends well for anyone involved.

Your Turn: Daggers Above

When was the last time that you got really angry at your boss for something that in retrospect was very minor? Think about it.

- What did they do to make you angry?
- What was the reason that you felt the way that you did?
- Was there a deeper root cause to your feelings such as frustration about your work or your career progression?
- How did you take steps to resolve the situation? Do you feel that your relationship improved as a result?
- What would you have done differently if your roles were reversed?

We've explored situations where you'll face scrutiny from your team and also when you will scrutinize others. It's a natural human trait, but how you deal with those feelings is important. Typically underneath bad feelings are other issues needing to be worked on, so as a manager you'll need to try and surface those from your staff. When you feel this way toward others, try to work with yourself to probe into why you're feeling that way.

Not all scrutiny and criticism needs exploring. Sometimes it needs deflecting or transforming. During stressful times, you may need to be the shield for your team. Although this may manifest in many different ways, we're going to look at two specific examples. You'll know how to:

- Handle times of *uncertainty and charged emotion* by looking at the concept of *wobble*, and the part that you play in it.
- Handle periods of *pressure* by looking at the concept of *the whip and the carrot.*

Throughout these scenarios you'll see where the political and emotional sides of management can come into play. Let's begin by looking at wobble.

Wobble

The higher up the org chart you go, the more that you'll find that the day-to-day concerns that you are dealing with are more abstract, uncertain, and just plain messy. If you're brand new to management, then dealing with this very human element of the workplace can be stressful. Whereas days spent as an individual contributor allowed constant focus around (mostly) well-defined pieces of work, your days spent as a manager open you up to issues and interactions that are much harder to define, contain, and resolve.

In addition to the unstructured and complex issues that you'll experience through managing humans, you'll also partake in discussions that have no clear correct answer. Should we hire this person or that person? How do we cut costs? How do we cancel this project and do the other one instead? Should we be building this product or that product? The higher the stakes, the more likely uncertainty and the increasing seniority of those involved will create tension, conflicts of personalities, and sometimes horrible arguments and falling outs. Yuck.

You'll begin to realize that part of your job as a manager is to shield your team from input that is too messy, disruptive, and emotional when it would be detrimental to them doing their work well. Often, you'll need to be the person that digests information and filters it down in a more palatable way. Sometimes you'll need to ensure they don't hear it at all. You need to *contain the wobble* that could affect your team.

Imagine, if you will, Jell-O molded in a conical shape with a wide base and thin, pointed top. This is your organization. Visualizing the Jell-O from the side profile, imagine an image of your org chart superimposed over it, with your C-level executives at the top and your individual contributors on their teams at the bottom, as shown in the image on page 194.

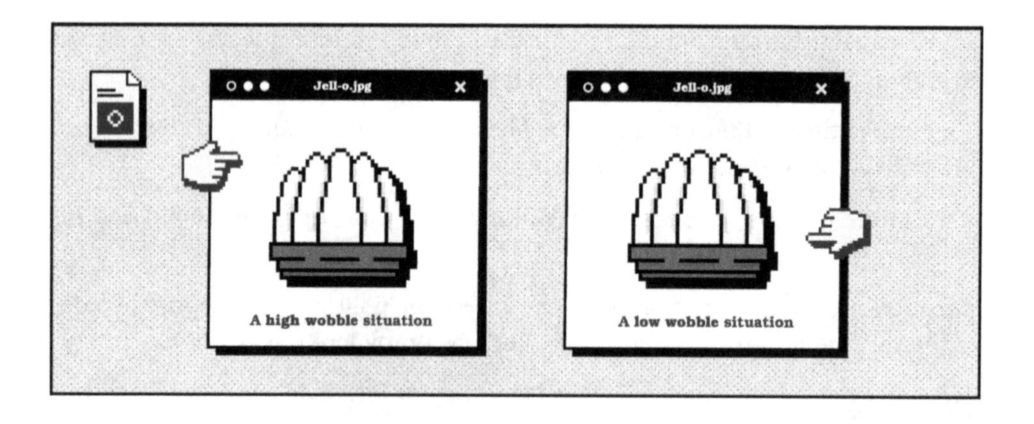

If there's a situation that blows up within one of the teams, such as a big argument over direction or stress and panic around a deadline, then imagine a finger giving the Jell-O a prod at the bottom of the mold. It wobbles a bit but settles again quickly. It's structurally sound down there. However, imagine the same thing happening at the VP or C-level. A big panic or fallout occurs and it *isn't* contained. Fallout ensues. Your finger gives the Jell-O a firm prod right at the top and the whole thing wobbles all over the place and takes a long time to settle. Do you see where I'm going here?

Not only do the people who perform the more senior roles potentially panic and explode over higher-stakes things (for example, the health of the company, the future roadmap, how to do a big re-org that might cause upset and redundancies) but they are also role models to the rest of the organization. Those that observe the leaders in an organization losing their level-headedness will look up to them and also panic. They won't know what issues they are panicking about, but they'll inevitably catastrophize and assume the worst. Rumors spread and suddenly everyone is uneasy and distracted. It can take a long time to settle: maybe weeks or months! Bad times.

Part of your job as a manager is to prevent this wobble from occurring. You must try your best to protect the part of the organization that reports in to you in times of adversity. This requires a good amount of emotional intelligence, judgment, and support. When a difficult or uncertain time requires a change in direction of your team, you want to be able to communicate this in a calm and reasoned way that results in those that report to you understanding the issue and being ready to work on change, rather than wanting to flip desks and set the whole place on fire.

Let's assume you're party to a discussion (well, argument) around pivoting one of the company's products. It just so happens that it's the product that your team builds. You've been invited to a meeting with the C-level leadership and there's a heated debate.

What should you do?

Listening and Observing

The first skill to practice is *mindfully listening and observing without judgment.* If a bad situation is unfolding, focus on your breathing, sit (or stand) straight and hear the other parties out. I'm aware this can be difficult if you have the board screaming at you, but do try. Take in the information. Take notes if that helps. Try to identify the parts of the situation that are fact and those that are emotion. Separate them. You'll want to focus on the facts in further communication and try to ignore the emotion, although identifying why these facts have caused high emotion is useful to think about.

Digesting

Since this situation involves your team, you'll want to *create some time to digest* between receiving this information and communicating it to them.

- What does it really mean to them once the emotion has been extracted from the situation?
- Does anything actually change?
- Will they need to stop what they are doing and do something else instead? How might they take the news?
- Are there any individuals that you think will react badly to this situation? If so, why is that?

Even if you think that you are the coolest cucumber, being in emotional situations will change your character temporarily. If possible, wait until some time has passed before delivering any challenging news. Contain the wobble until you've composed yourself. Going home and having some distractions and time to relax allows the subconscious mind to comb through the issue. Sleep is also great at calming the emotional metronome. Often, when waking up the next morning, nothing is as bad as it seems.

Communicating

Now it's time to deliver the news to your own people. You may want to *reframe the message* before passing it on. An example here would be one of your teams having their project cancelled because it isn't making enough money for the business. Your CEO and CTO may have locked heads and had an impassioned debate (let's face it: an argument), but as the team's manager you'll need to take that message and turn it into something more positive: that the team has done a great job, that you're grateful for their effort, that it was a challenging project for the business and sometimes these things don't work out, that most new ideas don't gain traction. You could introduce them to the cool new thing that they're going to be doing next.

You'll need to use your intuition to decide the best medium in which to deliver the news. Is it best talked through in one-to-ones before a broader announcement? Do you want to talk to the team as a group all at once? Do you want to hold off slightly before you know what they're doing next? Only you will have the answer. You know your team best. If you need advice, talk to your manager or a peer. The time taken to digest allows the right message to form and for you to communicate from a place of transparency and openness.

Never shy away from the facts, only try to reframe them for the greater good. People know when their manager is hiding the truth from them, and you'll be doing them a disservice by doing so.

Peer Support

You're only human after all, and emotion and change will take its toll on you. In *How to Win Friends and Influence People* you learned that building a network, especially with those outside of your department, can create a sounding board that can listen if you are dealing with difficult times, further softening the impact of the finger on the proverbial Jell-O. It gives you a wider perspective. Ensure that during tumultuous times that you are *looking after yourself* as well as doing your best to look after others. Talk to your manager. Talk to your peers. Talk to your family and friends. Then go and do something fun.

Your Turn: Think About Past Wobbles

Think back over your career to times when there were major wobbles at your company.

- What caused the wobble to happen?
- Did the wobble come after the news broke, or was it fueled by rumor?
- How long did it take for the company to get back to normal?
- What actions did the managers and leaders in the company take to stop the wobble from happening?
- If you had been at the center of the wobble, what would you have done differently for the impact of that event to have been less significant? Do you reckon that it could have been reframed in a more positive and transparent way?

The Whip and the Carrot

Work harder! Work faster!

Has anyone ever asked that of you and your team? How did it make you feel? Did it motivate or demotivate? Cause inspiration or decline?

Software engineering, like many other forms of technological and scientific work, can be opaque to those that are not skilled in it. Whereas you can watch a stonemason carve a statue, it can be difficult to observe how development work is progressing. "Just what exactly are all these people doing all day? They look pretty relaxed, right? Can't they just *work harder* and get this over the line quicker?"

Sigh. I'm sure you've been there. Let's pretend you're there again.

Let's imagine that the CEO pulls you aside and tells you that she thinks your team just isn't working hard enough. She rarely sees any of them staying as late as everyone else, nor getting in before them. She says that they spend a lot of time playing table tennis and visibly having fun together, but she feels they aren't producing enough output in return.

You know that this is a baseless conjecture, and it makes you boil over with rage. How on earth could they have this assumption when the reality is completely different? Let's try and explore their position by putting ourselves in their shoes and by being as empathetic as we can be.

Be objective and consider the other side of the argument: Why were you and your team asked to work harder or work faster? What is it exactly that the outside observer feels that your team is lacking?

Here are a number of reasons that you may hear as to why your team isn't "working hard enough." Typically it stems from a mismatch in what the external observer expects to see from your team compared to what is actually happening.

- *Lack of visible output.* Perhaps the team hasn't delivered anything that the observer has seen for a given period of time. This could be because the team has had a poorly defined project or has been subject to unrealistic expectations, bad luck, poor prioritization, and so on. Often it can be entirely not their fault. It can even be because they are doing a lot of behind-the-scenes work with infrastructure, technical debt, or refactoring.

- *Lack of "hustle."* The external observer may feel that there are particular characteristics that a team should be exhibiting, such as supposedly getting in early, going home late, being publicly present in communication channels, or some other manifestation of "working hard."
- *Lack of passion.* The external observer may feel that the team is doing incredibly important work, but the team is not motivated by it. Perhaps they find the project unexciting and it offers them little challenge and opportunities to learn.

Assuming that the person giving you the feedback about your team is important—and in this case we're talking about the CEO—you will need to think about how to handle this situation, depending on the truthfulness of their observations. You will either need to argue the case that it is not true, or if you feel that there is some truth to it, attempt to make some changes.

However, as we explored earlier in relation to *wobble*, you can't just let this filter through without some reframing, as it has a high probability of being damaging to the morale of the team. You also shouldn't just tell them off. You're much smarter than that.

You could imagine that there are two stick-like tools that you could use to make a team perform better.

- *The whip*, which is the metaphorical way of telling them to work harder and faster against their will.
- *The carrot*, dangled at the end of the stick, which makes them work harder because they feel motivated to do so.

You'll find that people external to your team, typically those outside of software development, will expect you to deploy the whip. However, being a good manager is about turning that whip into a carrot.

Let's explore this further.

The Whip

We can make some observations about software engineering that are true of many creative and scientific professions.

Firstly, the people that are on your team are working in an industry where *demand for talent dramatically outstrips supply*. If any of your engineers quit, they could very easily have multiple job offers within days. Given that this is the case, you cannot scare someone into working harder through making them fear for their job: they could just go out and get another one. Also, why would you want to be mean? You're so much better and smarter than that.

Engineers are also *self-motivated.* It's likely they're not doing this particular job because they have to. They're doing this job because they *want to.* What exactly motivates them can vary greatly from person to person: some enjoy optimizations to make things faster, some enjoy building customer-facing features, some just love problem solving. But the reasoning is all the same—the many years of difficult education and training to become a good engineer wasn't done through gritted teeth and poor conditions; it was done with a curiosity and passion. Getting too aggressive with the whip will get in the way of them doing a good job, which is what they are motivated to do, so they will inevitably leave.

Viewed through the lens of the in-demand, self-motivated engineer, you can see how when the whip is applied badly, with poorly defined logic, it inevitably causes problems for you as a manager. Setting fake deadlines is a terrible idea: they'll see through your deception. Telling them to work harder and faster with no clear reason or purpose will make you look stupid. Instead, you need to understand how to motivate in a positive way to produce the outcomes that you desire.

The Carrot

True leadership to increase throughput comes through fostering purpose and passion in your team. When your engineers are clear on their purpose in the organization and how they can move the dial for the company, they'll intrinsically perform better. When they are passionate about their work, they'll do a better job because they intrinsically enjoy doing it. Coaching a team toward this is much more difficult than just telling them to work harder and faster while being angry about it. It requires emotional intelligence, understanding how they work and what motivates them, and the ability to build the trust and rapport with them so that they want to join you on this journey. It's a long-term play, not a short-term play.

If your team isn't motivated by their work, or not operating as well as you'd like them to, then assuming that it's not down to performance problems, you'll need to foster the conditions under which they can guide themselves toward being a highly functional unit.

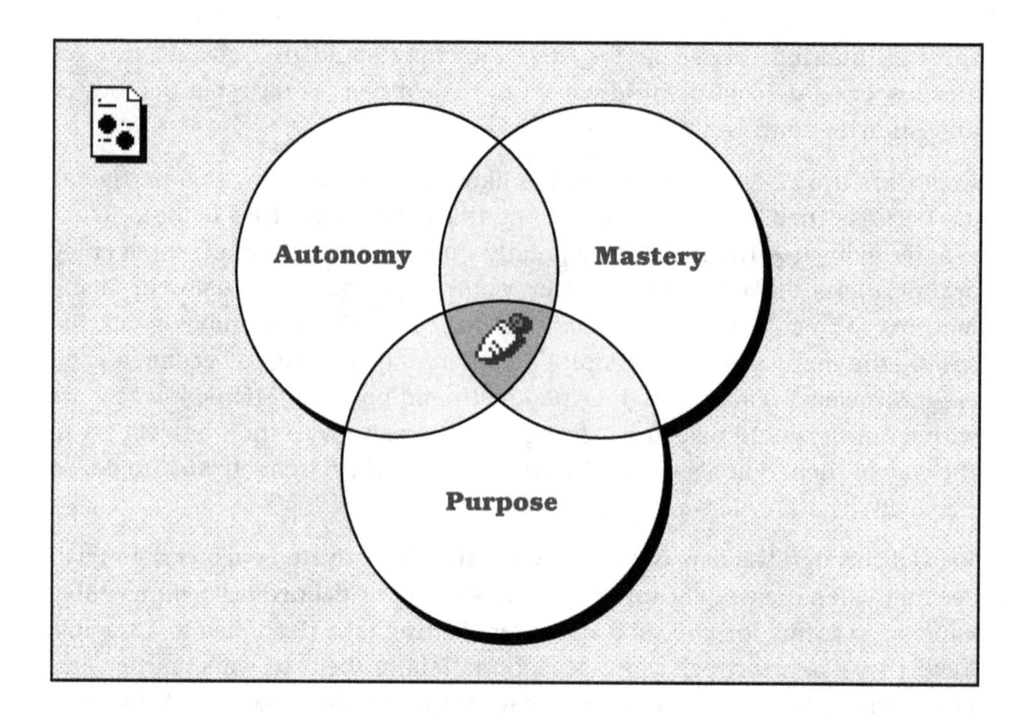

A useful model is that of *autonomy, mastery, and purpose [Pin09]*:

- *Autonomy*: We have a desire to be self-directed. Management as way of ensuring conformity and compliance runs *against* this principle. Instead, how can you create the conditions within your team to allow them to choose their own path while still working toward the collective goal? Can you delegate more to give your team more empowering work? Can you give them more choice over the way in which they solve problems and the tools they use? Can you let them approach problems in their own ways rather than having them take the same path that you take?
- *Mastery*: Everyone likes to improve their skills. It's incredibly satisfying. How can you delegate work within the team to ensure that your staff are in the zone of proximal development (see *The Right Job for the Person*)? How can you foster a culture of mentorship so that the team learns from each other while they are doing what is expected of them in their jobs? Are you regularly discussing their career development so that each day is a step toward a self-defined goal rather than just another dollar?
- *Purpose*: We want to contribute to something greater than ourselves. Are the projects that your team is responsible for clearly defined in terms of their purpose within the wider world? Are you able to create and continually refer to this message? For example, you're not just building database

infrastructure: you're allowing hundreds of thousands of people around the world to do their jobs better with your software. Surface the real difference that the team are making from within the work that they're doing so that they operate with a connection to why they're making the world a better place.

As you can see, working harder or working faster isn't what you should be focusing on. Instead, you should be creating the conditions that make your staff happy and productive through nurturing their autonomy, mastery, and purpose. Figuratively beating engineers with the whip to tell them to go faster and harder without reason is going to compromise the respect that you've earned from those that are working for you.

If you're facing external pressure about the performance of your team, then you need to start from within. Teams that operate while working toward the carrot of autonomy, mastery, and purpose will do better work, ship faster, and will have fun while doing so. No whips required.

Your Turn: Autonomy, Mastery, and Purpose

Consider your team and the concepts of autonomy, mastery, and purpose.

- Using the descriptions provided above, where would you rate your team on a scale of 1 to 10 in each of these areas?
- If you rated any of those areas poorly, then why is that?
- What do you think that you need to do to score them higher?
- Explain this concept to your team and get them to rate themselves. What do they think? How do their ratings align with yours?

Mount Stupid

Not only are you going to have to deal with people potentially picking at you and your team for a multitude of different reasons, you're also going to have to balance the multitude of people who want everything done perfectly yesterday. But to make this all even more exciting, you're going to deal with two incredibly fun concepts in psychology that will affect people's opinions and your ability to make decisions.

These are:

- *The Dunning-Kruger Effect*: A cognitive bias of illusory superiority in people of lower ability.

- *Impostor Syndrome*: A concept where high-achieving individuals are unable to internalize their achievements and fear being exposed as a fake or fraud.

These could be seen as the inverse of one another. Once you've learned about them, you'll spot this behavior happening everywhere. Let's look at them in more detail.

Dunning-Kruger Effect

A *1999 study [KD99]* by David Dunning and Justin Kruger presented the results of experiments that proved that, in many social and intellectual domains, people tend to hold overly favorable views of their abilities. Not only does this lead to poor decisions, it also means that people are unaware that they are making them! The paradox is that in those same areas, when the skills of the participants were improved, they were able to recognize the limitations of their abilities and therefore realize that particular decisions were bad.

What does this mean for us in management? Unfortunately, it means that there will be limitless situations where those who know the least about particular problems will feel the most bullish and comfortable with making a decision and will not realize that the decision is bad. This effect can be seen from both ends of the seniority scale.

- *Poor decisions by junior engineers*: High-achieving and confident junior members of staff, notably those who have just graduated with excellent grades, may not have the experience to make considered decisions about doing engineering in production. Their overconfidence in being able to solve a problem may give a project a green light, only for it all to not work further down the line. Their limited experience of production systems combined with their confidence in their own abilities can result in them not seeing any warning signs in their initial beliefs.

- *Rash decisions by senior staff*: They will be too far from the technical details to intuitively know whether a project is achievable or completely intractable. If they are confident, as your senior managers usually are, then they may be unable to be convinced that there isn't a simple solution that your team hasn't yet found, despite being told otherwise. This can lead to bad decisions about starting or stopping projects, outsourcing work, or blaming poor performance for a particular outcome.

The most dangerous place in the Dunning-Kruger effect was plotted to humorous effect by *Saturday Morning Breakfast Cereal.*[1]

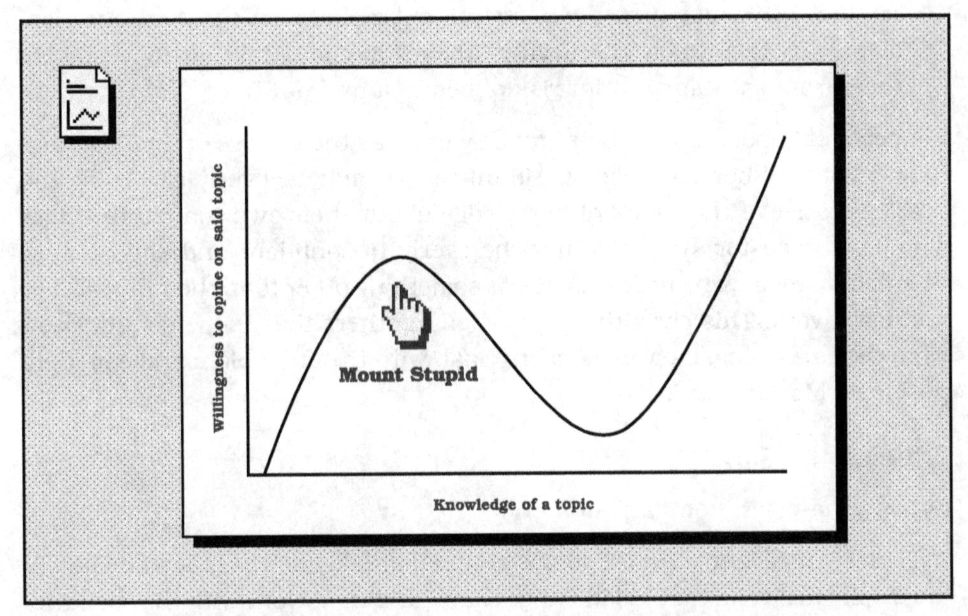

It's very easy to be on Mount Stupid, and the worst part about it is that you won't know that you're there. We can all experience it as it happens to all of us. Unfortunately, this can lead to a decision being made where nobody really knows anything at all as there is no counter-balance to an argument, or where the person with the most power, who happens to be on Mount Stupid, exercises their executive right, making everyone else feel dumb.

Impostor Syndrome

It's possible that high-achieving individuals may not give an outward appearance that is congruent with how good they actually are. This is called *impostor syndrome* and was noted in a *1978 study [CI78]*. This internal contradiction about one's ability can result in excellent individuals feeling like they are somehow a fraud—that they are faking it and will soon get found out. This leads to nervousness and lack of confidence in their own ability.

Like the Dunning-Kruger effect, it can play out at differing seniority levels:

- *Overly shy junior engineers*: High-achieving juniors may observe a lot of smart and senior folks around them and believe that they have absolutely no right to be there and that, in comparison, their ability is poor.

1. http://www.smbc-comics.com/?id=2475

- *Overly cautious senior staff*: They forget that many of the things that come totally naturally to them due to experience are through their hard work, not luck. This may turn them inward and make them cautious and risk-averse, as they know the devil is always in the details. It may prevent them from speaking out for risk of being "found out."

Elements of impostor syndrome can be viewed as the inverse of the Dunning-Kruger effect. Whereas those on Mount Stupid may be overly confident and brash because of their lack of knowledge about their own inabilities, those who have impostor syndrome may be overly unconfident and shy because they believe their own success is due to something other than their intelligence and hard work. This gives the impression to others that they may not know what they are doing, creating a nasty clash with the outspoken and confident people on Mount Stupid.

Bridging the Gap

So, as a manager, how can you help?

In short, you need to exercise your emotional intelligence to spot when people are displaying traits that fall in either camp, and then work with that behavior to help your staff overcome it.

Dealing with the Dunning-Kruger Effect

For junior members of staff being potentially reckless due to not knowing the ramifications of what they are doing, you need to show sensitivity. We've all been there. You definitely don't want to make anyone feel shot down by demonstrating your superior knowledge. Instead, you want to try and make junior staff come to the conclusion that they are being overconfident by themselves. Think of coaching (see *How to Win Friends and Influence People*): how can you keep the thought bubble over their head while they tackle a problem? Can you collaborate with them and subtly lead them to discover where the problem is much harder than they thought? Once they discover that a great number of technical problems are more complex than they initially seem, they'll have developed more mature techniques to analyze approaches and find the right balance of confidence and skepticism.

Senior staff who find themselves on Mount Stupid can often be difficult to deal with, because a senior person may be highly confident in a group; others present may by worried about arguing with them. Here you need to subject them to a similar process as that of junior staff, where they can arrive at the conclusion that something is harder than originally seemed to them. You need to pick your battles here: some personalities may be totally open for a reasoned debate or working it through together on a whiteboard, but others, depending on the situation, may benefit from discussion being taken offline, where some proper research can be done and results presented in a less confrontational format, such as an email or short document. Mount Stupid combined with big ego can cause awful verbal conversation. Instead, present only the facts; resist the urge to tell someone that they're being dumb. (Sometimes you'll have to surrender and leave a meeting in a haze of anger in order to revisit the facts later. It happens.)

Dealing with Impostor Syndrome

Brilliant junior staff who experience impostor syndrome need to experience repeated success to overcome how they feel. One way of doing this is by pairing them with a senior member of staff who is a good mentor and having them work through problems both abstractly and concretely, implementing them together through pair programming. We often forget that in education we're receiving specific grades and scores for the quality of our work and, over time, self-confidence can be fostered by seeing repeated positive results. The workplace doesn't offer frequent quantitative feedback like this, therefore regular interactions with more senior staff who can show the junior that they're doing a great job will build their confidence.

Senior staff experiencing impostor syndrome can forget just how much they know and can contribute. To reinforce their knowledge, pair them with junior staff. This actually benefits both parties, as shown in the previous paragraph. They will realize just how much they have to teach. For those that are reserved verbally, as a manager, try to weave them into debates by asking their opinion directly. "How's this problem playing out in your head right now? What are you thinking?" Make a concerted effort to bring their opinion into the room as often as possible. They'll soon see that their knowledge is valued, hard-earned, and useful.

The Dunning-Kruger effect and impostor syndrome play out in the workplace much more often than you think. Many people haven't heard of either. It's up to you to use your emotional intelligence to spot when your staff may be experiencing them, acknowledge that it's totally normal, and then try to employ some techniques to help your staff overcome them, without making anybody feel stupid or embarrassed.

Educating others about these concepts can also enable them to call you out when you're being brash about a decision you know little about or are being overly reserved when you actually have a lot to contribute.

Your Turn: Dunning-Kruger and Impostor Syndrome

It's time to think about the two psychological concepts that we just learned and to map them onto situations that have happened to you.

- When was the last time you were on Mount Stupid? How did you get off of it? Did you get called out for being wrong, or did you come to that conclusion yourself?
- In your new management role, does impostor syndrome describe some of the feelings that you've felt so far?
- When was the last time you observed either of these phenomena in your colleagues? What was the situation and what do you think caused it?
- Go through the concepts with your staff in your one-to-ones. It's fun. Can they think of any times when they experienced or observed this behavior?

It Isn't Just Humans...

Well, you've made it through a venerable feast of concepts and ideas that may cause some very frustrating days as a manager. But you're well equipped to deal with those challenges.

Here's what you've learned:

- That you will face *scrutiny and judgment*, and you will feel it toward others as well.
- That bad situations create *wobble*, and that the higher up in the org chart that they occur, the worse the effects on everyone else.
- That others may want you to apply the *whip* to make your team work harder and faster; however, you need to reframe that into the carrot that makes your team autonomously drive themselves forward.
- That we're all subject to the *Dunning-Kruger effect* and *impostor syndrome*. You've learned how to spot it and what to do in those situations.

Now, if the humans being hard wasn't enough, then guess what? The projects themselves can be hard too. Let's look at that in the next chapter.

I love deadlines. I love the whooshing noise they make as they go by.

➤ Douglas Adams

CHAPTER 11

Projects Are Hard

"Argh! What do we do?"

OK, so today isn't going so well.

The marketing launch for your feature is tomorrow. You've known that for a long time. The team has known for months. In fact, it was them that set the deadline. The team has done everything right. They've nailed every sprint, they've had stakeholders using it from the start, and they've shipped almost daily to production. There was 99.9% confidence that everything was going to be absolutely fine on launch day.

But it isn't fine. Far from it.

It all came out of the woodwork at the last minute. Services somehow unable to talk to each other intermittently over the network. Spurious and confusing errors in the logs. Yesterday, data corruption. The team deserves so much better. Why now, one day before launch? Why them? Why you?

You feel the heat. You feel the accountability. The responsibility. What do you do?

You decide to send a message to product marketing.

```
You: Hey, are you there?
Alex: Yep, what's up?
You: I think we need to delay, there's some major bugs
Alex: What?! We can't do that now
Alex: The press release is out, the video done...
Alex: Can't the team just work harder and get it over the line?
```

You get that familiar feeling of a tight jaw when you read something that really triggers you. Work *harder*? If only they understood... But has the team actually worked hard enough? You question your own performance. You doubt yourself. No, no, that can't be true.

You send a message to your product owner.

```
You: Hey, have you got a sec?
Jo: Yeah - all good for tomorrow?
You: Not really
Jo: What's going on?
You: Loads of bugs. Struggling. Might need to delay.
Jo: We can't!!
Jo: How did this happen? Surely this was an easy feature?
You: Huh?
Jo: This sorta stuff should have been easy to build.
```

Your jaw gets even tighter. You clench your teeth. Another non-developer with a strong opinion about development. Do they realize they have no idea what they're talking about? You notice that your right hand is curled into a fist on the table. You breathe deeply. You count to ten.

"Right folks, let's look at the logs again."

As a manager, you're going to deal with some stressful times. It's part of the job. No person is perfect. No team is perfect. No project is perfect. However, no matter how high your tolerance or your inner calm, you're going to go through periods that really push you to your breaking point.

This chapter is all about those stressful times and how to deal with them. Being a manager means being accountable during the good times and the bad times. But especially the bad times. Your ability to deal with *bad* situations is your true test as a manager. And there will be plenty of bad situations. There will be late projects, unexpected scope changes, and pressure, pressure, pressure. But try not to worry. Help is at hand.

Here's what we're going to look at in this chapter:

- *The Eye of Sauron*: When the deadlines are looming, it feels like the entire company is looking at you and your team. We will see what you should do when the Eye of Sauron casts its gaze upon you, and what to do after it's gone.

- *Victims of your own success*: As companies succeed and get bigger, it gets more challenging to get work done at the same pace. We'll look at why and what you can do about it.

- *Scope, resources, and time*: When you feel the pinch, these are the three levers that you can pull. We'll find out how.

Don't let this chapter put you off from being a manager. It's going to make you a better one. You can get mentally prepared during peacetime so that you make better decisions during wartime. Remember—when it does get bad, this chapter is going to always be here to help you.

Let's get going.

The Eye of Sauron

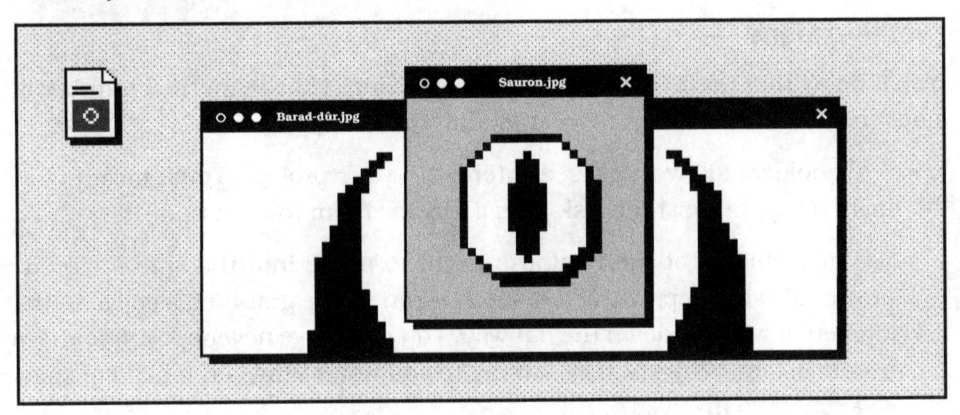

Even before getting into management, I'm sure you'll be able to recall the moments that you were under intense pressure to deliver.

Sometimes this pressure can come from *anticipation*: your team just happens to be responsible for delivering the most important new feature for the company this year. It's being announced on stage at an event that's already booked. Argh! Pressure to deliver can also come from *catastrophe*: parts of your infrastructure may not have scaled as expected and are continually on fire, and unless a new solution is developed, your customers are going to go elsewhere.

In these moments, you'll have felt what I like to describe as the gaze of the *Eye of Sauron*. Yes, that flaming, piercing stare of the Dark Lord from the top of Barad-dûr in Mordor. (If you have no idea what I'm talking about, have you never read or watched *Lord of the Rings*? Go and do that immediately!) Whichever way you turn, the entire business is looking toward you. It's uncomfortable. You can feel the heat. There are emails, chat messages, ticket comments, interruptions in person, you name it—it's constant and stressful.

- "Is it looking OK for next week?"
- "Our clients are asking for an update. How's it going?"
- "Is it fixed yet?"
- "Can we not put more people on this?"

Depending on your mindset, you can turn these tough situations into a challenging but rewarding experience for your team, or conversely, you can totally fumble. Handled correctly, you'll be looking at career growth. Handled poorly, and you may find the next high-stakes project goes to another team instead.

Warning Signs

You'll know that the Eye has turned its gaze onto your team by a number of cues that begin to become more frequent and intense.

- Stakeholders show increased interest in your project. You'll have to bat them away, rather than ask repeatedly for them to turn up.

- Senior members of the business begin to probe into the status of your project at every opportunity, such as when you're grabbing a coffee in the kitchen or walking down the hallway. You may have never spoken to these people before. Why are they talking to you now? Why is the head of sales so keen to be involved?

- Your boss, or boss's boss, is being more direct and intense with the progress of your work. Why do they care more than usual?

- You're noticing how your upcoming feature is being hyped internally and externally. It may now be perceived as the headline launch of the year, even though that was not apparent when you started the project. Why is the business sending out animated teaser tweets when the UI isn't even designed yet?

- Or, quite simply, everything is on fire and the platform won't work unless your team digs their way out of this hole. The support ticket queue is getting bigger, and bigger, and bigger...

Regardless of how the situation has unfolded, it's important to increase your vigilance and take extra effort to manage your and your team's way through this period of heightened pressure. Handled deftly, you'll fully own the tough situation and have something to celebrate once you deliver. You'll also earn the team some much needed breathing room afterward.

Under the Gaze

OK, it's crunch time. The deadline is next week. It's looking 50-50. What do you do? Are you feeling the heat? Although the increased pressure will make your job harder, there are some principles that can help you through these intense and difficult periods. Try to ensure that you're applying them daily, through your discussions, meetings, and decisions.

- *Align the team*: When you're under the Eye, your team will probably know. But if they don't, or if you're purposely shielding them from knowing, then that's a bad situation to be in. Utilize the pressure in a positive way: align the team around what they need to achieve, make sure everyone understands how to succeed, and then facilitate them moving toward the goal.

- *Over-communicate*: At times of immense pressure, you'll want to increase the visibility of what the team is working on. On top of your usual way of demoing your work, consider writing a weekly (or more frequent) update to stakeholders. You could even record a short video if it's quicker. Depending on the size of your company, a weekly newsletter to the wider organization might be suitable too. Either way, you'll want to make it absolutely clear what you're working on, how you're progressing, and any key decisions that you've had to make.

- *Invite responses and feedback*: Frustration can be prevented if you open up a clear channel of communication for others to use. Make a chat channel, mailing list, or similar so that curious and restless minds have a place to interact.

- *Release frequently*: Since your cadence is of utmost importance, ensure that you're releasing as frequently as possible so that your stakeholders can follow along with your latest builds. The more time that you have for feedback in stressful situations, the better. Don't keep code held back until the deadline; it just makes the event more stressful and the resulting mega-merge and big-bang release might cause all sorts of bugs. Use feature toggles, keep shipping to production, and tell people where they can look at the new builds. High-performing teams continually ship.

- *Be pragmatic*: As dates loom nearer, or as the system continues to ignite, you'll need to make pragmatic calls on speed of development, quality of the code, and creation of technical debt. As much as it can be painful for idealists in your team, you'll likely end up shipping some shonky code to get the work over the line. However, make note of every hack you put in so that you can tidy up and refactor later.

- *Lead from the front*: As a leader, you need to set the example for the rest of the team. Put in the work. The hardest projects can become career-defining moments. Own them and be there.

A successful high-stakes project can be fantastic for you and your team: you'll bond through difficult times, further strengthen your trust and rapport with each other, and you'll have something big to celebrate.

When the Gaze Is Averted

Let's assume the pain has passed and you're in the aftermath of the marketing launch for your new feature. Retweets are pinging off everywhere, the company blog posts are churning out, and you can hear the salespeople ringing the bell again and again and again.

However, what you do next is very important for the morale of your team. Consider this: Glastonbury Festival takes place on a dairy farm, and occasionally the land will have *fallow years* that allow the ground to recover after all of those cows and people have been traipsing all over it. You need to do the same with your team, even though they haven't had to endure people dancing on them and being eaten by cattle, although it may feel that way after a deadline.

Here's how you can round off a tough project and allow some time to recharge:

- *Celebrate*: This is one of the most critical things. The team has worked extra hard and they've met their commitment. Take them out for lunch, or drinks, get some cakes shipped out to the office, put on a gaming night—whatever makes them happy. Make sure that you say thank you for what they've done.

- *Tidy up and clear down technical debt*: As the deadline approached, a whole bunch of little shortcuts may have been taken: a hack here, a missed unit test there. Put aside the next sprint to refactor and tidy up at a more leisurely pace, while fixing any production bugs if they arise.

- *Do self-guided project time*: Time can also be put aside in the coming weeks for some self-directed learning. Allow the team time to experiment and to learn something new. This change of pace and direction puts some mental space between the last project and the next one.

- *Reflect*: Arrange a *project retrospective [DLS06]* meeting. It's a focused way of reflecting on the whole process and discussing what could be handled better next time. Even if the project was run perfectly, the project retrospective is an opportunity to mentally close the current book before opening up the new one.

- *Plan and regroup*: Take some time to think about what's next. Are there any initial explorations that can be picked up? Is it time for a *design sprint [KZK16]*? It's time to start the discussion, to think about some initial planning, and to get excited about the future.

It's safe to say that—like any period of crunch—intense periods of work are not sustainable. If a team finds themselves under the Eye too often, it'll cause burnout and attrition. There may be greater problems with the company if it's stress after stress after stress.

This may be unavoidable at a startup, but that's an exceptional circumstance: those that are there are fully committed to the challenging ride. But at a larger company, it's important to consider how projects and expectations can be managed to allow different teams to feel the heat of the Eye in an alternating sequence as time progresses, allowing the other teams to temporarily shift down a gear and regroup.

For example, it's advantageous for your marketing department to space out product launches to maintain a steady flow throughout the year. Why not take advantage of this? Your product organization and your engineers can balance these intense periods of contraction with periods of release and recuperation.

As a manager, you should also actively fight for after-project space for your teams if the business doesn't give them the opportunity by default. When a big project is over, push back on demands to create the room for the rejuvenating activities in the previous section.

It's a marathon, not a sprint, after all.

Your Turn: The Eye of Sauron

When did you experience your last immensely stressful project? What was it?

- During the project, what did your manager do to make the project better for the team? Did it help?
- If you were the manager in that situation, what would you do differently using the examples in this section?
- Once the project was over, what did the team do next? Did they have an appropriate fallow period?
- Next time your own team goes through a critical project or period of intense stress, make the case for them having time to celebrate, regroup, and recharge.

Victims of Your Own Success

If you've ever worked in a small company that has grown into a big company, then you may have experienced the feeling that things have slowed down as you've gotten bigger: the time it takes to make decisions, the time it takes to ship new features, and the time it takes to merge code and for it to be deployed to production.

The *Mythical Man Month [Bro95]* had this covered way back in 1975 with a concept that became known as Brooks' Law: that "*adding manpower to a late software project makes it later.*" In fact, this observation can be made more general, insofar that adding more personpower to a team, department, or company makes everything more complicated. Knowledge work cannot be easily split into discrete tasks: there's always communication involved. You can't chop a programming project into two without increasing the amount of communication required to get it done. So even if your team, department, or company has gotten bigger than it was before, you can't expect a linear return on investment for the number of people added.

Even though this concept has been around for longer than many people have been on the planet, it still comes up time and time again:

- People within the business may be frustrated with the progress from Engineering. After all, you're three times as big now, so why aren't you doing three times the work?
- Your users may feel like you've slowed down as you've become more successful. You used to ship twelve big increments a year. What happened?

It's Not People, It's Productivity

As manager, you should ensure that arguments around speed do *not* become ad hominem. Although people are quick to point out that those at a startup may be working ridiculously long hours, that simply isn't sustainable or desirable at a bigger company. It should be possible to get everything that you need done and still have a life outside of work. A few individuals are not making the entire department or company slow. It's a natural side effect of growth.

It's about being a victim of your own success. It's about having to deal with legacy systems. It's about the increased communication overhead of having ever more people. The more accurate way to phrase your perceived slowness is that the *productivity per head* is decreasing, and this is the problem that needs addressing.

You'll find that if you're successful:

- *Business as usual gets harder.* Doing well creates more work. Customers expect better SLAs. The security of your applications needs more work. Your monitoring and alerting needs improving. Oh, and that on-call rota needs setting up. You'll have to scale that storage better. Will it need rearchitecting? Just keeping the lights on takes an increasing amount of time and effort.
- *You're dealing with more and more legacy code.* Yesterday's feature is today's technical debt. A new product direction or company pivot can introduce dirty hacks and rewrites. Even if you don't introduce crazy hacks, the ever-increasing size of your codebase makes it take longer to work out how best to introduce new functionality, then implement it.
- *Communication and process overhead are ever-increasing.* Startups can make unilateral decisions at speed. Large multinationals can make decisions at a glacial pace. How many people are required for consensus? How many meetings and emails does it take to get there? *Communication channels [Bro95]* far outgrow the number of staff: there are *n(n-1)/2* potential channels. A 400-person company has 79,800 of them!

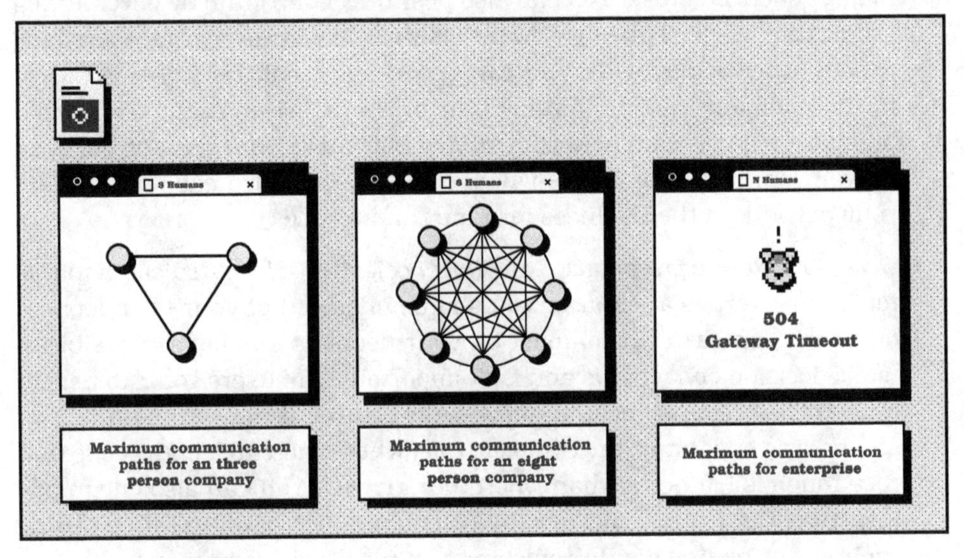

The fact that productivity per head decreases as a company gets larger has *allowed startups to flourish [Chr16].* That feature that a piece of large enterprise software can't decide on or deliver quickly because of the inherent complexity of steering a large ship becomes a gap in the market that a nimble startup can begin to undercut.

So What Can You Do?

In addition to being a shield for the team so that they aren't affected directly by arguments over speed, you have to accept the fact that productivity per head will decrease as the company gets larger, and you'll have to be able to explain it to them. Perhaps you can buy them a copy of *The Mythical Man Month* for Christmas. However, as a manager, you can encourage three behaviors from your team.

You can:

- *Expose the hidden complexity.* Dealing with an increasingly bigger system with an increasing number of people makes even adding the simplest features harder than others may expect. As a manager you can make sure that the hidden details are exposed. For example, let's say that a salesperson just wants a new auth integration added. Easy, right? Well, maybe not. Let them know what that really entails: dealing with logging out user sessions after some period of time, two-factor authentication, ability to change their password and username, and the list goes on. You can help build empathy for what it means to do things to a high level of quality.

- *Always show progress.* You can also positively contribute by never letting people assume you're being slow by ensuring they've always got something to see, hear, or play with. You can encourage a culture where everyone in the company can be informed on progress if they want. You could publish a regular team newsletter with what you've been working on, open up your sprint demos to the whole company via video call, or lean into your network in the business for input and advice.

- *Develop software pragmatically. Whole books [HT00]* are dedicated to this topic; however, as a manager you should ensure that your team focuses on having a measurable impact on your users as quickly as possible to make decisions on what is and isn't important. Are users going to benefit from this work in some way, now or in future? If not, don't do it. You should also encourage the *campsite rule*, where each area of the code that they touch is left better than when they arrived. You can also encourage investment in future efficiency by reserving a percentage of time during each sprint or timebox to removing technical debt (why not try 20%?). Also, never be afraid of deleting features. Did it not ship fully? Delete it. Did interactions with that area of the product wane significantly over time? Make the hard call and get rid of it. It will only slow you down in the future.

So remember it's not that *people* get slower as a company gets bigger; it's that *productivity per head decreases* when those staff are part of a larger organization. As a manager, it's important for you to make sure that your teams are developing software in the most pragmatic way possible and that requirements, progress, and achievements are transparent to the rest of the business. The knock-on effect is that those who question your team's progress can engage with you in conversations that criticize the *right* things—which we'll consider in the next section—rather than the wrong things (for instance, not "working hard enough" or performance of individuals), as the right things can be collaboratively discussed and altered where necessary, without judgment.

Your Turn: Productivity per Head

Think back over your career. Consider the following:

- Which point in time did you feel that you were most productive? What were you working on and with how many people?
- Conversely, have you ever worked on a project or at a company where it was impossible to get anything done? Why was that?
- How do you think that you can keep the productivity per head in your team high? What sort of changes might you need to make in the team?
- Ask your team how productive they feel in your next one-to-ones. How do they feel that this is trending over time?

Scope, Resources, and Time

Back in *Join Us!* we considered the Triple Constraint with regards to hiring. Now we're going to use it with reference to your projects. If you want something good and fast, it'll be expensive. If you want something good and cheap, it'll be slow. If you want something fast and cheap, it'll be poor quality.

As a manager, you'll have a finite number of engineers at any given time, and you will routinely find yourself juggling your ongoing projects with the need to forge ahead with the next great thing that you should be building.

When discussing upcoming projects, you may face conflicting opinions from different departments, including your own. These particular examples are taken to an extreme—in reality people are more pragmatic—but for the sake of discussion we present the following opinions in the light of a hypothetical new project appearing on the horizon.

- *Commercial wants it now!* Your salespeople are finding it difficult to pitch against your competitors because of features that are lacking in your application. This next feature can't come soon enough, because their confidence in you is paramount to their success.

- *Product wants it all!* Your product manager has a grand vision for the roadmap for the next year. The designs look beautiful and the offering compelling, but you can see that it represents a huge amount of work. Probably too much work.

- *Engineering wants it right!* Your engineers see this new roadmap as the key trigger to redesign a large part of the system. Building it around what already exists will introduce a lot of technical debt that will cause serious pain in the future, so they want to rewrite it all.

Three different opinions that are all equally valid. A healthy tension exists between them: where do you begin the discussion?

Now it goes without saying that you would not want to compromise on quality. I would argue that if you're happy with shipping poor-quality software, then you're probably in the wrong job.

Instead, you have three levers that you can adjust to find the right compromise with new projects. They are scope, resources, and time. Sometimes you will have flexibility over all of them, and sometimes you won't.

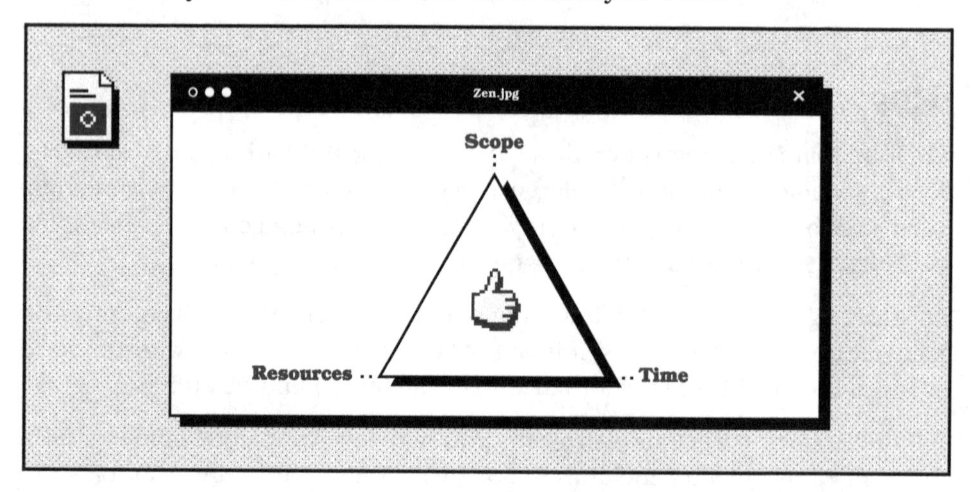

- *Scope*: The definition of what the project is going to deliver.
- *Resources*: The number of engineers that you are going to assign.
- *Time*: The duration that you have to work on it.

Imagine that you're beginning to explore a new project. How do you frame it among everything else that is going on?

Scope

Let's dig into scope. Your engagement point is to have your team work with your product owner to break the product or feature into epics and stories that each deliver tangible added value to your application. This is your backlog. Then, from there, work together to prioritize it.

By being an active, rather than passive, participant in backlog prioritization, you can ask important questions around the scope of the project. If you're working to a deadline, then a bit of work up front can save your team a lot of pain down the line.

- *Encourage categorization of features into must, should, could, and won't.* Work labeled as "must" absolutely needs to happen. Anything with the "should" label is important, but you could probably get away without it for the initial milestone or deadline. For example, it could be less time-critical and could ship as a first increment once all of the musts have been delivered. Those labeled "could" are desirable but not necessary. The "won't" category notes that something has been thought of and consciously decided against. Labeling your work in these ways makes it absolutely clear that if the team is up against it, you can just commit to getting the "musts" done and defer the other tasks until later. This prioritization technique, as shown in the chart on page 220, is called the MoSCoW method.[1]
- *Encourage stretch goals.* After labeling the features and ranking them in order of prioritization, work with your product owner and team to decide where the milestones are. Should the team aim for all of the musts along with a handful of shoulds for the first release? Which should-haves could be part of the second release? Deciding these up front can save a lot of future stress and panic by giving you buffers to work with. Knowing which parts of the scope can be dropped in case of delays and technical issues gives a safety net for when things go awry.

A little extra thought up front can make projects run much more smoothly. If you're in a pinch, *you can use the lever of scope to exclude any work that hasn't been categorized as a must-have.* You could even revisit those categories

1. https://en.wikipedia.org/wiki/MoSCoW_method

as a project progresses. It might be the case that some of those musts aren't that necessary any more. Revisit, refine, and adapt.

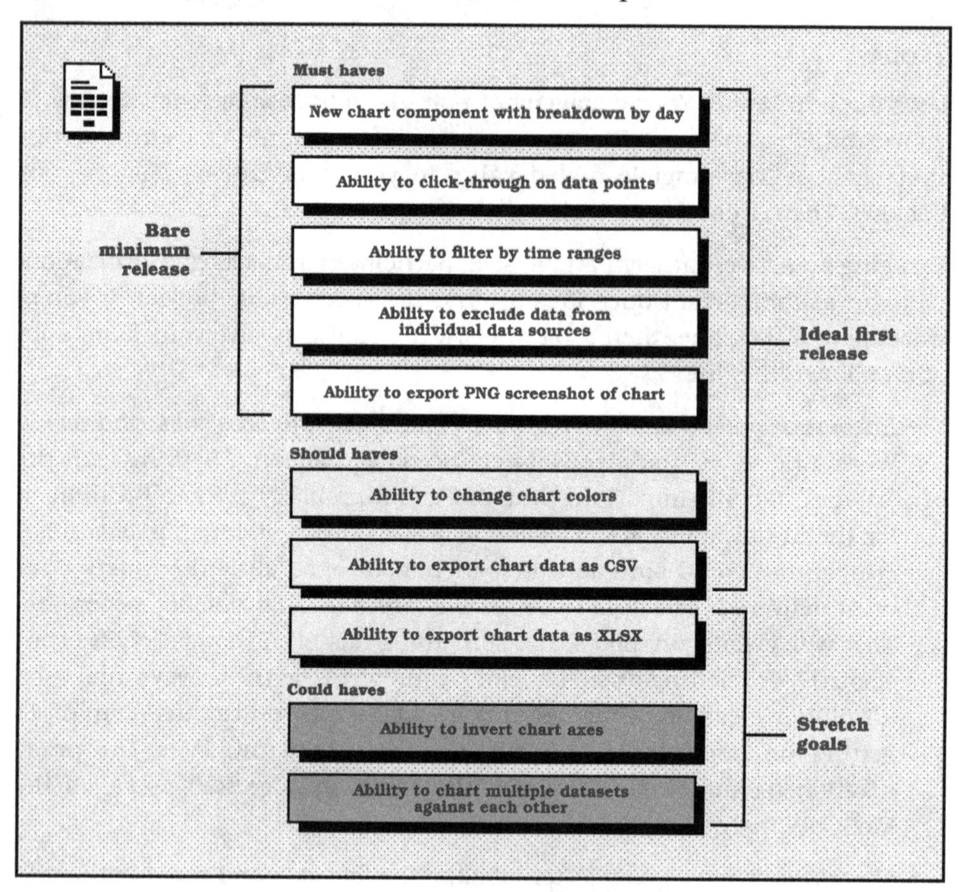

Resources

What size of team is going to be required to get this work done? It can be tempting to throw the proverbial kitchen sink at an important new project, but each part of that project has a number of considerations:

- *Task parallelism*: How much of the work in the scope is sequential, and how much can be done in parallel? Two separate services can be developed independently, but iterations of the same service or component cannot.
- *Code vicinity*: The last thing you want is multiple engineers all working on top of each other in the same part of the code, causing nasty merge conflicts with their commits. You don't need four spanners scrabbling over the same nut.

- *Technical difficulty*: Is this looking like a business-as-usual increment of functionality in your existing application, or is this a new innovative piece of architecture? This drives decisions about which of your staff might be required to work on it, or at least offer their time to oversee it.

Everything is always a trade-off. As we know that adding more people to a slow project has the *effect of making it slower [Bro95]*, you should categorize each piece of work as outlined previously. Which pieces can be developed in parallel, and which sequentially? Can any parts be worked on by multiple engineers at the same time? If not, could they pair program on it?

If the proverbial really does hit the fan, moving people around *isn't your strongest lever to pull*. Nothing is ever as easy as just adding more people. However, if you've categorized the work, you should know which pieces have the *best chance* of being sped up. This is when you can ask your manager, with confidence, whether you could borrow an engineer to get a particular parallelizable piece of work done, and know that a critical sequential piece is only going to be sped up by drinking ever more coffee.

Time

You cannot bend time. If you can, then you probably don't need anything you're learning in this book. But, given that you probably can't, what can you do if it looks like you're going to be running out of time?

- *Investigate that deadline*. Work with your product owner, project manager, or whoever decided the deadline to see what it *really* means to let it slip. Is it that nothing is actually affected other than the email campaign going out a bit later, or is the CEO going to be unveiling the finished project on stage in front of thousands of people? If it's the latter, remember that even Apple will reveal something on stage at WWDC and say that it's coming in a few months. Dig deep for what will *really* happen if the date moves. More often than not, it's possible to shift it.

- *If all else fails, pull your other levers*. If the deadline can't be moved, then you'll need to revisit scope and resources. It's either that or acquiring Bernard's watch. (*Bernard's Watch* was a British children's drama about a boy with a pocket watch that, when pressed, would stop time but leave him unaffected. It's the most prized artifact in software engineering.)

You can't create more time, but you can use it wisely in decisions. As Andy Hunt says, time is neither created nor destroyed, merely allocated.

Make Your Levers Transparent

If you have a considered approach to your new and ongoing projects using the levers of scope, resources, and time, then you can make clear choices that are easy to explain when questioned about your priorities. Making your decisions and their reasoning visible internally to your department can also help teams see how their projects fit into the bigger picture.

The trade-offs of scope, resources, and time guide debate around pragmatic concepts that you can discuss openly and honestly, shifting the conversation toward how further trade-offs can be made when under pressure, rather than inviting ungrounded criticism about how hard you and your teams should be working instead.

Visibility of these trade-offs also empowers your teams to help each other too: one team may offer to lend a hand because they know another team's project is critical and immovable and their own more flexible.

Your Turn: Scope, Resources, and Time

Think about projects that you have been part of over the past few years, both the good and bad.

- Did you have any projects where engineering, product, and commercial severely clashed in their priorities? How did it get resolved?
- Did you experience any projects where scope, resources, and time were discussed throughout the project? Who led those discussions and how did they help?
- Have you discussed scope, resources, and time with your team with regards to their current work? Bring it up in your one-to-ones. Do any further conversations or changes need to happen as a result?

And Relax...

And there we go. Two whole chapters about how both humans and projects are hard. Are you regretting your job choice yet? I hope not. You can do this. Here's what you learned:

- That pressure and looming deadlines can feel like you're being stared at by *the Eye of Sauron*. You learned strategies for detecting it, coping with it, and most importantly, how to recover afterward.
- That any company can be *a victim of its own success*. As more people join a company, everything gets more complicated. We studied how this is linked to productivity per head.

- That skillful manipulation of the levers of *scope, resources, and time* will get you out of all sorts of tricky situations and you can now use them like a pro.

Whew! Tough, isn't it? Next up we're going to look at one of the most important commodities of being a manager: information. How should you deal with secrets? How do you communicate just enough information? And what are workplace politics all about anyway? Let's find out.

Reports that say that something hasn't happened are always interesting to me, because as we know, there are known knowns; there are things we know we know. We also know there are known unknowns; that is to say, we know there are some things we do not know. But there are also unknown unknowns—the ones we don't know we don't know.

> *Donald Rumsfeld*

CHAPTER 12

The Information Stock Exchange

"Did you hear?"

You spin round on your chair. "Hear what?"

It's Ben.

"About our wonderful CTO."

"Erm, I don't think so? What might I have heard about her?"

Ben looks toward the ground. "Well, there's a rumor going round, but it's just a rumor."

"What is it?"

"That she leaked customer details to another company."

"What?"

"Yep. All of our customer details including their search logs."

"No way. I can't believe it."

"Nor can I," says Ben.

"I feel really uncomfortable about this," you say.

"I think more people need to know," says Ben. "I think you need to tell the team."

"How do we even know that this is true?"

"Let's just say that I know. Trust me."

You think again about what was said. It sounds awful. Surely people have the right to know if something like this has been happening. But then again, what good would that serve? They'll just ask you the same questions. Where did this information come from? When did it happen? How do you even know that it's true? Should you trust Ben? But why would he tell you this? What would he gain from telling you? Your head spins. You feel conflicted and confused.

"Are you OK? You look like you've seen a ghost!"

It's the CTO.

You feel a cold flush as everything Ben shared with you plays back in your mind.

"Uh, yeah. Hey! I'm fine. I'm fine."

"Are you sure?"

"Yes! Yes. All good."

"OK, just thought I'd ask. See you in a bit."

"Yes! Bye!"

Wow, that was close. She seemed pretty normal. Almost as if nothing had happened. Did it even happen at all?

As a manager, information is your currency. The more you know, the better the decisions are that you can make. But not all information is good. Information can be gossip, rumor, or malicious lies. The longer that you spend in this role, the more that you'll be exposed to. This chapter is all about how to properly handle information: the good, the bad, and the confidential.

You're going to explore the following:

- *Spies and gatekeepers.* We'll look at differing motivations for collecting and discovering information and see which is most appropriate for you as a manager.
- *How to share just enough information.* Building on the previous section, we'll look at various types of information, see how to categorize them, and then see how to consistently share just enough.
- *Workplace politics.* We'll then turn to this common phrase. What does it mean? Is it good or bad? Should you be getting involved?

Are you ready to get going? Bring your trench coat, hat, and sunglasses. We'll rendezvous at midnight.

Spies and Gatekeepers

Take a moment and consider a piece of confidential information. What comes to mind? A redacted dossier leaked from the FBI? The inside of Area 51? The salary information of your direct reports? As a manager, you're going to have access to all sorts of sensitive information, whether you like it or not. In your one-to-ones, you'll talk about confidential topics, professional and personal. In addition to this, due to your position as a manager, others may divulge secrets to you as they know that they can come to you in confidence. Or worse, they may spread rumors to attempt to curry favor.

Before we dive into learning about how to share just enough information and how to handle workplace politics, let's look at two concepts related to information in order to glean some insights into how *you* should think about the information that you collect as a manager.

Let's start with a sexy topic: espionage. Spies are cool, right? James Bond escapes the exploding train via the window with merely seconds to spare, clutching the laptop that's storing the nuclear launch code. Ethan Hunt scales the Burj Khalifa while being battered by heavy crosswinds. But what actually *is* spying? According to Wikipedia, espionage is the act of obtaining secret or confidential information or divulging of the same without the permission of the holder of the information. It's a high-risk, high-reward activity.

Although you may not find yourself abseiling down the Hoover Dam or hacking into the mainframe of the CIA any time soon, as a manager you *do* have some similarities with spies, insofar that you do obtain and handle sensitive information that is of the utmost importance. It just might not be as important as the nuclear codes. Although it may seem bizarre to compare ourselves with spies, by doing so we can highlight some important points:

- We should never attempt to obtain information without the permission of the holder of the information.
- We should never attempt to obtain information for the benefit of another party.
- We should never attempt to obtain information using deceitful means.

But hang on, doesn't all of this sound entirely obvious? Well, yes. Perhaps it does when you read it. But isn't it tempting sometimes? Once I was accidentally sent a spreadsheet from Finance containing the salary information of everyone in the engineering department, rather than the filtered version showing only my teams. What did I do? I told them what they'd done and deleted the spreadsheet without opening it. I didn't have permission to see that, despite having been sent it. You may find yourself in similar situations, and you'll need to remember *not* to be a spy.

The inverse is also true. People may try to extract information from you for their own gain or for other's gain. Be careful of conversations after work at the bar. Be aware of others who may try to pry to find out various pieces of secondary information that they can piece together to infer what they really want to know. Just be cautious.

OK, enough negatives. Is there a better model that you can use to understand how you should operate with information? Fortunately there is. Let's consider the process of *gatekeeping* in communication. Much like a gatekeeper would control access to the gate of an ancient city, gatekeeping is the action of filtering information for further distribution. Typically it represents how information is filtered by the media before it is presented to the public. This happens at many levels, from journalists picking exactly which aspects of stories to cover, through to particular purposeful biases that selected media outlets may have. Stories may be selected due to their impact, content, familiarity to the public, and their proximity (think local versus global news).

Although you may not be running a global media conglomerate, you do have access to information. You do choose how and when to broadcast that information and to whom. *As a manager, you need to act as gatekeeper.* You have to:

- Decide *what* information should be heard by which parties.
- Decide *when* that information should be shared.
- Decide *how* that information should be framed.

You'll see this play out in all manner of situations in varying levels of impact. You may have to choose the right time and the right way to tell someone that you're going to be letting them go. You'll have to pick the right moment and message to let the team know that their project is being paused so they can work on something that the company deems more important. You'll have to consider how best to handle the message that the business has had a terrible year and it's looking likely that there are going to be redundancies.

How should you operate as a gatekeeper? No situation or piece of information is the same. This is why you need rules. In the next section we will look at how to share just enough information, but before we get stuck into that, there are two rules that you should abide by when it comes to dealing with *all* information. Both are simple and both have been around longer than we've been on this planet.

- *The Hippocratic Oath*: an oath of ethics historically taken by doctors, shortened here to the following succinct phrase: *first, do no harm.*
- *The Golden Rule*: an ethic that is the basis of a number of religions and cultures: *treat others as you would like to be treated yourself.*

When handling any information, if you remember the rules above, you'll get most of the way there. But let's take a deep dive into the nuances of handling information.

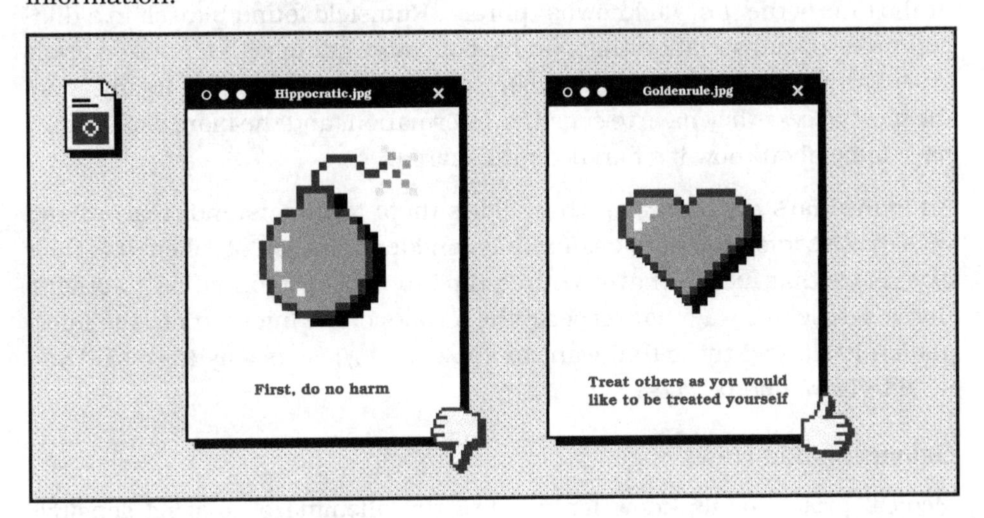

How to Share Just Enough Information

On February 12, 2002, U.S. Secretary of Defense Donald Rumsfeld stood in front of journalists and the media at another U.S. Department of Defense news briefing.

Facing another probing question about the lack of evidence to link the Iraqi government with the provision of weapons of mass destruction to terrorist groups such as Al-Qaeda, Rumsfeld began his reply, with little idea that he was about to coin the phrase that he would be remembered by.

"...because as we know, there are known knowns; there are things we know we know. We also know there are known unknowns; that is to say, we know there are some things we do not know. But there are also unknown unknowns—the ones we don't know we don't know."

I remember watching the briefing on BBC News. My initial reaction was that Rumsfeld had said something tautological and utterly ridiculous, but, in retrospect, it has been labeled a smart distillation of complex matters. I think I now agree.

Although the public would declare this phrase as a Rumsfeld creation, *his memoir [Rum11]* mentioned that it was commonly used inside NASA, of which Rumsfeld likely heard a variant of when he worked on the assessment of ballistic missile threats to the United States in partnership with William Graham, an administrator at the space agency.

At the time of the "known knowns" phrase, Rumsfeld found himself in a difficult professional situation that can become more acute with increased levels of seniority. Generally, the more senior that an individual is in an organization, the more access they have to sensitive information, and the more careful they have to be about how it's handled and shared.

An individual's *experience* is what allows them to understand, reason, and vet sensitive information to ensure that confidentiality isn't broken. It is their *experience* that means that when it's time to deliver information to others, that it's done in a way that respects the owners of the information, the information itself, and those that want to know more. This is why it's a skill you need to *learn and practice.*

Delivering Bad News

Medical professionals know far too well the dilemma of sharing sensitive information. When communicating with their patients, trust is established through openness and honesty. If a patient has been diagnosed with a fatal illness, then the delivery of that information must be done transparently, sensitively, and kindly.

This requires a great deal of knowledge and understanding on the part of the physician, both in terms of how to summarize and present the information, but equally importantly, how to deliver it in a humane way with empathy and candor. Ethics are also important to consider, as the physician must also understand how to handle delicate situational intricacies.

For example, consider how fatality policy in hospitals requires the next of kin to perform the initial identification of the deceased. This may mean that a close relation may be refused to see the deceased until the next of kin has done so—an ethically difficult quandary.

Furthermore, is it wrong to withhold the specifics of a diagnosis, even when it isn't life threatening, from someone who is suffering from serious mental health problems and therefore could be exposed to more risk as a result of knowing the truth? What if there's no concrete reason to withhold information of a diagnosis, but instead their family is requesting it be kept secret from them?

Maybe we should be glad we're in software.

Trends Toward Transparency

As a manager, you'll be required to make regular decisions about how much you should share with other staff and when. The easiest option with any sensitive subject is to not say anything at all.

But is keeping everything a secret by default the right thing to do? Definitely not. Unless there's a critical reason for hiding information, it should be shared, although care should be taken in how the message is delivered.

Earlier in the book (*Interfacing with Humans*), we explored the foundations of the relationships that you want to build with your staff. Modern management is about facilitation, empathy, openness, and candidness.

Some startups in our industry have been taking radical steps toward openness in their culture. Buffer publish their staff and salary information in public for the entire world to see.[1] They also publish their salary calculator, which estimates how much you would earn if you worked for them in a particular role in a given location.[2] The idea is that when everything is on display, there's nothing to hide, therefore transparency brings consistency and fairness.

Revisiting our medical information–sharing dilemmas, that industry has long since changed its default stance to openness. In a survey in 1961, 10% of physicians believed it was correct to tell a patient the exact details of a fatal cancer diagnosis, a percentage which had *changed to 97% by 1979 [SFKI16]*. More recently, the NHS is implementing trials of genomic tests to predict your likelihood of fatal illness in the future.[3]

In work, in life, and in health, we want transparency. But how can we do that while still respecting that not every detail can be shared?

As we touched on previously, as your time spent as a manager increases, so does the exposure to sensitive information. But what sort of sensitive information

1. https://bit.ly/2P3oJVs
2. https://buffer.com/salary
3. https://www.bbc.co.uk/news/uk-47013914

are we talking about? You'll be party to some or all of this information during your time as a manager. Some of it is information *about* the team, and some of it is information that may signify something about to happen *to* the team:

- *Compensation.* You'll know the salary data of your direct reports, including any inconsistencies between them. You may know that Jim is overpaid because he negotiated well when joining, so Alice is underpaid in comparison. You need to normalize that over time.
- *Performance issues.* You'll know whether any of your staff is underperforming. For example, you could have one member of staff on a PIP, and that should remain a secret to the rest of the team.
- *Wider company changes.* Perhaps there's a re-org being planned that hasn't been finalized yet. It affects your team but isn't yet ready to communicate.
- *Redundancies.* A bad year for the company may mean that some people will need to be let go soon. That list of staff is confidential.

This is a privileged set of information, and it's your duty to ensure that it is treated with the utmost respect. For issues that you can resolve on your own, this is usually fine. You can keep a secret, right? But it becomes more difficult when you need to call upon others to assist you with reasoning about, or making decisions with, the sensitive information. You need to begin to share it with others. But how do you know what to share, with whom, and how should you act as a gatekeeper to ensure that you follow the Hippocratic Oath and the Golden Rule?

Consistently Just Enough

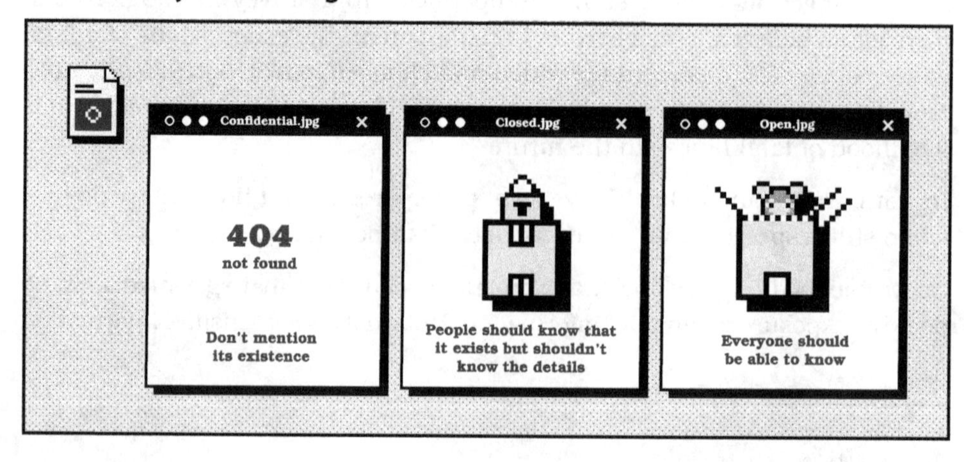

We can think about how to share sensitive information by trying to categorize it further. Perhaps we could sort it into three categories:

- *Completely confidential*: Aside from those that have been given authority to know, nobody else should. For example, a round of redundancies would fall into this category. This information is not shared.
- *Closed box*: Think of it like a present under the Christmas tree. You all know it's there, but you don't know what's in it. The process or concept isn't sensitive, but the contents are. For example, people will know that pay reviews are being done, but they won't know each other's pay, nor will individuals know until the process is finished. This information requires you to *filter* and *gatekeep* it.
- *Open box*: This information isn't sensitive at all, such as which staff are in each team. This information can be shared widely.

Across these categories, the following rules apply:

- You are *consistent* with how you treat information with different people.
- You always share *just enough*.

Even though these two principles are straightforward, it's surprising how easily they go wrong, and very rarely through malice.

Being Consistent

We fight our biases. When it comes to sharing information, you need to ensure that you're aware of what it means to be human. From an early age, humans have an innate desire to share with others. Babies point, and toddlers pick up items so that you look at them and see what they are seeing. When you're given sensitive information, sometimes you have this unexplainable desire to share it with others. You may be compelled for many reasons to share something with someone at work. You may know somebody extremely well and feel a duty to keep them informed, even if it's not relevant to them. You may even subconsciously be trying to build rapport with someone by letting them into a secret.

Regardless of your intention or relationship with someone, you need to be consistent. You should only share information with a person if:

- It is relevant to them and their job.
- They have a reason to know.
- It benefits you both, such as it unlocks further conversations.
- You only share what they need to know.
- They can be trusted to keep any sensitive information a secret.

Otherwise, you should question your motives for telling them in the first place. Only you will know the exact answer to who should be told what and who shouldn't. But you should be aware of the variables that can affect your decisions and the way in which to share that information.

Your Turn: Sharing Conundrums

Consider the following situations. What would you do and why?

- You have placed one of your team members on a PIP. You know that your other staff are frustrated when working with that person. Should you tell them you have done this to show that you're addressing the problem?
- A round of redundancies are coming up. You know that your team isn't affected. Should you tell them ahead of time not to worry?
- You find out from another manager that the company is going to get acquired by Apple. Should you tell anyone? How should you find out more?
- You find out that the budget is there this year to give everyone in your team a 20% pay increase. Should you share the news with your colleague, who is a fellow manager of another team?

The Meaning of Just Enough

So what does sharing just enough actually mean? It's straightforward for two of the three information categories. *Completely confidential* information should be just that. *Open box* information should be broadcast as much as is useful for everyone's knowledge.

The trick is getting the closed box category right. My own take is that the *existence* of closed box information should be broadcast as much as is useful for everyone's knowledge, except that the full details are not disclosed.

Many people default to closed box information being treated as confidential, but this can be perceived negatively. For example, if the pay review process is underway, why not regularly update everyone that it is progressing, even though you're not going to disclose the details until a later date?

Keeping silent in situations where just enough information can be shared can make staff feel as if there's a reason that you're not talking about it. Often, after rumor and gossip, that reason can become negative (for example, "they're not doing pay reviews this year!") when the real reason can be quite positive (for example, more time and money is being spent on competitive benchmarking).

In the absence of information, people tend to assume the worst. So try not to let that information be absent in the first place. Classify the information that you hold, and make sure that you share just enough of it so that staff feel included and well-informed.

If you don't, your known unknowns might be linked to weapons of mass destruction, and we all know how that turned out.

Your Turn: Classify This!

Classify the following information as completely confidential, open box, or closed box to your team. How should you handle that information, and how should you share it with your team, if at all?

- The salary information of each of your staff.
- The fact that you are planning your team's budget for next year.
- The fact that you are planning your team's budget, and one of them is going to be made redundant.
- The company-wide inflationary increase for salaries at the end of the year.
- The agreed percentage salary increase for top performers.
- What your team is going to be working on next.
- That your team's current project is getting canceled. You know what the next project is.
- That your team's current project is getting canceled. You don't know what the next project is.

Leave Nobody Behind

Assuming you're handling your information well, then the task doesn't quite finish there. As a manager, you also have a duty to keep people informed. Leave nobody behind.

You should be continually trending toward consistency with the information that you share: ensure that everyone has seen the same version of the same information. Has an important decision been made? Summarize it in an email. Have various people expressed concerns directly to you? Answer them publicly in a place where the whole team can see them. Ensure that you mix private interactions and public broadcasts so that no person is left behind.

Keeping People in the Loop

Here are some strategies for keeping your team in the loop with pertinent information.

- You could send a weekly digest email to the team of any interesting and relevant information that you've been working with. For example, this could be planning hires for the coming year. Invite them in to observe what you're doing.
- Briefly mention what you are working on outside of your ticketed work at your stand-ups. For example, you could mention you are writing reviews or working with your manager on end-of-year promotions and salary adjustments. Keep them in the know.
- Give yourself fact-finding missions! Speak to your network, find out what's going on at the periphery of the business and within other teams. Write up what you've discovered on your travels for your staff.

Workplace Politics

We've looked at the intricacies of handling information. However, the information that you handle is just a small subset of all of the information that is flowing around the company at any given time.

The word *politics* is defined by Wikipedia as the way that people living in groups make decisions. Building software is all about making decisions: what to build, how to build it, who should build it, how to market it and sell it, and also what not to build as a result.

However, the word *politics* itself probably doesn't make you think of a political scientist observing behaviors in a local community. Instead, I assume you think of campaign trails, elections, and political parties. We are united and divided by differing ideologies. Companies have their own internal ideologies and culture that encode how they treat people, make decisions, and get things done.

Work, as in life, is filled with many independent actors in a complex system. Each actor is motivated by differing interests, passions, and values. Each is working independently toward the greater good of the company, and politics are what arise from the continual negotiation, persuasion, and debate of ideas.

The term *workplace politics* has often been used to describe purely negative situations such as those being taken advantage of, proverbial backstabbing, or people being poorly treated for the gain of others. I would say that is a subset of workplace politics: it can potentially happen, and it does, but it's not the wave that you should be looking to ride. You can use politics for good.

Politics in the workplace are always going to happen. You shouldn't choose to *not* engage with them, because if you don't, then you'll find that your career suffers as a result, especially as you spend more time in management. Instead, the art is to understand the politics of the workplace to discover how you can navigate them and use them to your advantage while at the same time acting for the greater good. Ego, power, unwritten rules, and implicit culture unmask themselves at higher levels of an organization, and you need to be able to harness them with a clean reputation and, even better, use them for the benefit of everyone else.

Let's dig deeper into what they mean.

How Do Politics Arise?

Politics typically arise because of tension between different types of social structures:

- *The org chart.* This is the most obvious place where politics can occur. It could be debate between you and your boss about your own interests, or conflict between you and your direct reports. It also manifests in individuals reaching down or up through the org chart to promote their ideas and increase their influence.
- *Close-knit informal groups.* There will always be groups of people who are close to each other and protect each other through friendship, camaraderie, or shared interests. They have no formal power structure in the org chart, but they lobby and work together.
- *Influential people.* Singular influential people, who may have much more say than others because of their tenure, celebrity, or bargaining power can make situations political or difficult to navigate because of the difficulty of building consensus without them.

As you spend more time in a managerial role, you'll see how decision-making is rarely an easy task. You have to navigate the structures and people just described in order to make sure that things move forward in a constructive manner.

Like in real politics, all political situations and negotiations involve an element of risk. How should you conduct yourself among the various groups lobbying for various, sometimes contradictory things? How does your team fit within all of this?

Getting involved in the wrong type of politics can be harmful, so you need to be able to protect yourself from the kinds of interactions that are toxic and only exist for those involved to be cliquey, spiteful, and malicious. However, getting politics right builds your authority, influence, and ability to get things done, opening up further doors in your managerial career.

Using Politics Positively

Let's explore ways that you should use politics to your advantage. The word *advantage* is loaded. But what we mean is finding ways to positively increase your output as a manager, ensuring your team is well connected and exposed to meaningful and impactful work, and having a positive effect on the rest of the department and company.

Connecting with Teams and Groups

Who makes decisions in your organization? Who is influential? Who are the close groups of individuals that think similarly, and who are the rival factions? Identifying this up front allows you to navigate sensibly through the political landscape and gives you the best chance of knowing who to go to to build consensus on particular issues, and who to approach differently or even avoid.

Take note of the different teams and divisions. What are their priorities? What motivates them? What do they care about and what are they knowledgeable about? Which individuals have sizable influence and why? Is it because of their tenure or their technical prowess? By mapping out the organization and learning how it currently works, you can identify which groups and individuals you can collaborate with on different issues with the least amount of friction.

Once you're done, set yourself a challenge to introduce yourself to them and begin building some connections. Remember that to have influence and to make an impact, you'll need to *win hearts and minds*. You're connecting with people to meet your colleagues, offer them your support, understand them better, and help them get things done, and vice versa. Positivity and kindness prevail.

Building Consensus

As companies grow in size, projects and initiatives move forward through collective effort rather than just the force of will of an individual. When working, you'll need to understand that consensus—at least as much as you can get—is important. In small companies and startups you can just take the proverbial bull by the horns and do whatever you want alone with few repercussions, but larger companies are different.

You can start small and informal. Let's use an example. If your team wants to do something dramatically different to the codebase for their next project, then it's important to take as many people as you can along for the journey at the same time. It's likely that your team won't be the first that has thought of doing something like you're proposing, so start by having some informal conversations with those that are senior, influential, and close to the matters at hand.

Assuming that informal conversations have been successful, then you can announce more widely that you'd like to try a proof of concept pull request, or even just write an idea paper for circulation. It's important that those you initially talked to are able to sponsor your efforts and offer their support, and that any work that you propose is just that: a proposal. Build consensus by making others feel like they always have the opportunity to contribute to what you are suggesting rather than it appearing to be a mandate. It will unlock the ability to make wider-reaching decisions.

Being Yourself

Appearances and interactions are important. You need to be yourself in a consistent manner to ensure that you're able to engage well with others and represent your teams correctly and respectfully. Don't pretend to be somebody you're not. Don't act a part to attempt to impress. Just be you. That's more than enough. You're great.

Always be open, transparent, respectfully critical, and clear on where you stand. Have the confidence to be open to being proved wrong and to be accepting if you are. Be open to *disagree and commit* to initiatives. Never push agendas for the sake of serving only one's self. Fundamentally your means of conduct comes down to the Golden Rule: treat others how you would wish to be treated yourself, and set the bar high.

Additionally, remember that when you become a manager, whether it's right or not, your position in the org chart grants you more power and you must wield this power respectfully and wisely. In the game of politics, you must make sure that your relationships with particular staff are never seen to be favorable or unfavorable for personal reasons, otherwise you can be seen to be cliquey or nepotistic and this will make you less trustworthy. It may also bite you in the future. You need to engage equally and fairly with all, regardless of your personal relationships with them.

Disagree and Commit

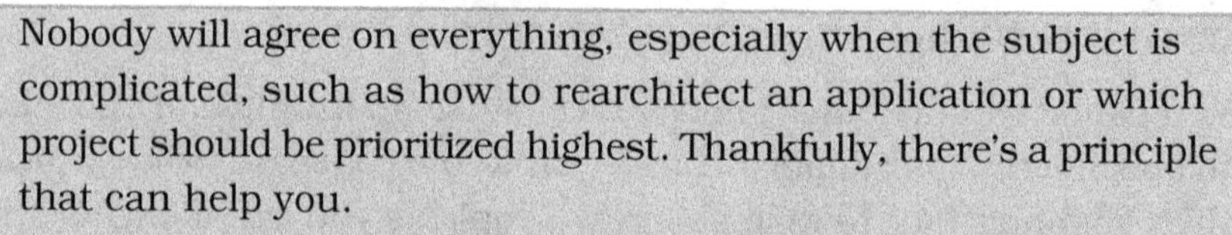

Nobody will agree on everything, especially when the subject is complicated, such as how to rearchitect an application or which project should be prioritized highest. Thankfully, there's a principle that can help you.

To *disagree and commit* is to state that a group hasn't achieved consensus on a decision that's being made, but once the decision is finalized, they'll commit to following it. This avoids the lack of consensus carrying over beyond the decision boundary, preventing progress.

Building Your Network

Previously we've talked about building a network of peers from different areas of the business to share information, get feedback, and sound out your ideas. In workplace politics, your network of peers is important as it allows you to be more broadly informed about how the wider business feels about your own initiatives and priorities. It also gives you a chance to trial ideas before taking them any further, allowing you to initially operate in a safe, cross-disciplinary setting. Continually foster this network and use it to make yourself a better manager.

Using Politics Negatively

We've explored a number of ways in which the politics of the workplace can be used positively to both your advantage and the advantage of everyone else. That's awesome. But there are many negative ways that you can engage in workplace politics that will at best result in conflict and at worst have a negative impact on your career. You don't want to do these. Let's look at them.

Misuse of Power

While your new-found managerial seniority may allow you to just tell people what to do, without winning hearts and minds, you'll gradually erode your respect in the eyes of your peers and ultimately your influence. I like to imagine each manager having an energy bar, like in a video game, that is depleted when a controversial override on a decision is made or an unpopular direct order is issued. You must use overrides tactically and sparingly. The bar replenishes when you move forward with your team in a congruent manner where they are motivated to go on the same journey as you.

Level-Jumping

I'm sure we've all been guilty of going over someone's head to their manager because it's quicker and easier than going via the chain of command. Now, there may be times that this is quick, easy, and convenient. But the person being left out in the middle feels *awful*. Perhaps things are going on in their team, division, or department that they are none the wiser about. Sometimes you may have a good relationship with your manager's manager or even be friends outside of work, but you need to make sure that you go via the proper channels and bring your own manager along for the ride. Otherwise they'll feel like somebody else is influencing their destiny behind their backs. (An exception to this rule is skip-level meetings, where a manager will have occasional check-ins with their direct report's staff to see how they're doing.)

The same is also true for those who run a large organization with many layers of management. If you're going directly to the direct reports of your own direct reports about issues that should really involve the person in between, then you're meddling and it shows disrespect for the person being left out. Could you not delegate this to them? Why not? Could you not coach them to do so?

Being Unprofessional

I'm sure that this goes without saying. If you wish to be influential and effective as a manager, then you need to engage with others professionally, respectfully, and kindly. This may be in contrast to how you feel in particular situations, especially if discussions are emotionally loaded, but you owe it to all that you interact with to be open, honest, transparent, and without ulterior motives. You need to operate for the greater good.

Those who are unprofessional erode the trust that the organization has in them, and in higher levels of the org chart where trust is of utmost importance, those who cannot demonstrate it will find themselves unable to progress their careers. Karma always comes back around.

Going Rogue

If you think that the best way to demonstrate how good your idea or initiative is by just doing it without anyone knowing, then it's likely to cause much more conflict down the line than if you'd built consensus in the first place. Going rogue is similar to building up technical debt: it gets worse the longer that it continues and it's harder to back out from.

For example, if you thought the best way to make big architectural change in the codebase was to do it silently rather than taking everyone else along for the ride, then you may find yourself unpopular when those controversial code changes end up being forced through because of an impending deadline for a feature, or even worse, blocking the whole deliverable.

If you're unable to build consensus, then maybe your idea or initiative isn't as good as you originally thought it was. Use the opinions of others to balance your own views and reveal your biases. You'll be better for it.

Personal Gain

The last negative area should come as no surprise: you should not use workplace politics for malicious personal gain. Bending ears to force through your candidate against the will of others so that you get the referral bonus is a terrible and unethical thing to do. Using your influence to force others into situations they don't want to be in for your own gain is also seriously bad juju. Abusing your position to make others feel small, powerless, or marginalized is the worst form of politicking and will always catch up with you. Just don't. Ever. You are here to serve others, not yourself.

It's Time to Decompress

We've had a trio of heavy chapters. We've explored a myriad of ways that work can be stressful: humans can be hard, and so can the work itself. Then, in this chapter we've covered:

- *Spies and gatekeepers.* You've learned that you need to be a gatekeeper, rather than a spy, even though that means less about jumping from planes and more about keeping others informed with what they need to know, when they need to know it.

- *How to share just enough information.* You learned how to categorize information and how to share just enough with others so that you respectfully keep them informed while keeping the confidential information confidential.

- *Workplace politics.* We looked at the ways in which you can use politics for the good of yourself, your team, and others. We also highlighted the bad stuff you should be steering away from.

Among the humans, the work, and the information, you can feel like a ship that's continually battered by the wind and waves. You fight to keep control. However, for your well-being and mental health, you need to let go. You need to embrace the full catastrophe of everything you can't control. We'll look at that in the next chapter. Get the kettle on, turn down the lights, and get comfortable. It's time to look after yourself.

You can't stop the waves, but you can learn to surf.

> *Jon Kabat-Zinn*

CHAPTER 13

Letting Go of Control

It's the weekend. Saturday morning, to be precise. It's time to relax, unwind, and de-stress. At least, it's meant to be. You're lacking sleep. Your first coffee couldn't come soon enough.

You spent your Friday night staring into space over dinner while you worried about Monday's deadline. You ended up checking through all of the in-flight development tickets and reading the latest comments. Why hadn't Ben replied to any of your requests for status updates? Does he not feel the same pressure that you do?

Later that evening, as you lay in bed, it wasn't your book that you held in your hand, it was your phone. You were refreshing your email inbox again and again and again even though nobody on your team would be sending any emails at that time of night. But you couldn't stop yourself. You needed to know whether everything was going to be alright. The itch needed scratching, but it was always out of reach.

Instead of getting that much-needed rest, you were wired. Was everything going to be ready on time? Was everyone else aware that this work really needed to get done? A restless night's sleep followed. You dreamed of lines of code, of infinite errors while your code was compiling, and of being chased down the corridors of your old school by a man-eating bug ticket while the lockers flapped open and shot calendar pages in your face.

Now it's the morning. You can't stop thinking about last week. Did you have all of the conversations that you needed to? Did you make it clear what was needed? Was there anything that you could have communicated better? Would it benefit you to drop a bunch of emails into your team's inboxes so that they get them first thing on Monday morning?

How are you meant to stop thinking about work?

We live in a world that bombards us with information, both inside the office and in our free time. Push notifications, emails, adverts, beeps, boops, and blips. As you will have learned by now, being a manager opens the door to more context switching, interruptions, and channels of communication than you would have experienced as an individual contributor. With time, you can feel attached to the input. You need it to feel like you're doing a good job. It hangs around in the evenings and the weekends. It can override the desire to cook proper meals. It can prevent you from lacing up your running shoes.

As a manager, you do your work through other people. Those on your team and those that you influence can do more work than you can ever do, but they're going to do it in a different way than how you'd have done it yourself. During times of stress, the urge to take back control and begin meddling and micromanaging is there. But you shouldn't succumb to it. It's easy to yearn for absolute control, but it will eat you alive.

This chapter is about letting go. It's about putting space between yourself and your managerial work and proving that by doing so, paradoxically, it will make your work better.

Here's what we're going to learn:

- *How to let go of tasks.* You'll drive yourself crazy if you get involved in the details of every little thing that your team does. You'll drive yourself even crazier if you stress about every eventuality. We'll revisit why it's beneficial for your output for you to let go, then look to the Stoics for how to frame your relationship with your work so that you worry less.

- *How to let go of your preconceptions for effective work.* As much as keeping on top of every notification, meeting, and email is comforting and even addictive, it might be seriously messing with your brain. We'll explore why. We'll also look at how you should carve out time for yourself as a manager to unlock your creative thoughts, and why you should be stricter with your inputs.

- *How to let go of work after work.* To get better at your job, you'll need to look after yourself. We'll have a look at techniques for caring for your mind and body so that you can recharge every day and come back the next as a better manager.

We'll begin by showing you just how essential letting go is for your success. But guess what? You're already doing it.

Transcending Tasks

Let's remind ourselves of our favorite equation. Your output is defined as:

The output of your team + The output of others that you influence

Now, consider both parts of that equation. Who is actually doing the work? Well, with regard to the output of your team, it's certainly not all you. Each member of staff on your team is working toward the team's collective output. With regard to the others that you influence, those people aren't on your team to begin with, and you might not even know the details of what exact work they're doing! *You work through others.* That's your job. It's what effective managers do.

Earlier in the book in *Interfacing with Humans*, you learned about delegation. We covered how delegation is the tool that you use to give just the right level of support for each task such that it gets done to the correct standard. In *The Right Job for the Person* you saw how keeping your staff in their zone of proximal development via delegation continually increases the skill of the person doing the task.

In the same chapter, you learned about autonomy, mastery, and purpose. Mapping the right tasks to the right people and delegating well contributes to autonomy (since they do the work in a self-guided manner in a way of their choosing) and mastery (since the work allows them to practice their skills and improve them).

Isn't it interesting that letting go of control can be beneficial in so many ways? Your staff improve their skills by stepping up to challenging work, you increase your output, the team improves, and the company gets better. Quadruple win.

So, you already know about letting go, don't you? You might as well just skip this chapter. Or should you? Stay with us for a minute. Despite being well-versed in the tools that you require to sufficiently let go of control of each individual task, it's sometimes not as straightforward as it seems.

In your practice as a manager so far, have you felt any of the following?

- How *stressful* times such as deadlines or disasters can stir a deep desire to take delegated work back and just do it yourself because you know that you can definitely do it quickly and to a high standard?

- *Anxiety* about the status of a delegated task, especially when it's before or after work and you can't contact the person that is working on it? Has the ticket not been updated for a number of days and you're feeling that tight-chested panic?
- *Frustration* that some work hasn't been done to the standard you expect, but that was because you didn't have the time to offer the right level of support? Has that frustration ever turned to *anger*?
- Have any of the feelings above happened to seep into your personal life, giving you a short temper or bad mood around your friends or family when doing something that should be simple and enjoyable such as making a coffee or picking up groceries?

It happens. Delegation has a dark side. When doing your job to a high standard involves letting go of so many tasks, even though you have the framework for how, you cannot escape the fact that you're a human and you'll want everything to go exactly as you want it to, even though this desire is impossible to satiate. The impossibility of that desire can make your emotions begin to eat you alive. You'll have restless nights. You'll get angry when the water tank hasn't been refilled in the coffee machine. A minor inconvenience can become a major one.

Trusting others and trusting your plans is challenging. You'll doubt them, you'll doubt yourself, and you'll worry about every outcome. So what can you do? Well, we're going to take some inspiration from some ancient Stoic philosophers.

"What?" you say. What indeed.

Stoicism and You

"But I didn't get into technology to learn about philosophy," you cry. "That course was so boring!"

Well, let me try to convince you otherwise. Stoicism is a branch of Hellenistic philosophy, founded by Zeno of Citium in the early third century BCE. (The Hellenistic period, which was heavily influenced by ancient Greek tradition, spans from the death of Alexander the Great in 323 BCE to the rise of the Roman Empire and its conquest of Egypt in 30 BCE.) Rather than being a lofty and academic philosophy, it's more a practical, logical, and ethical philosophy of life. Despite its age, it is still applicable today. The Stoic philosophers believed and taught that we should live our lives in the present moment, in accordance with nature, and use our logic and reasoning to act in a virtuous way that contributes toward maintaining our tranquility. The Stoics used the term *tranquility* to describe what we typically call *peace of mind*.

Fundamentally, the Stoics wanted us humans to eradicate our negative emotions. They wanted us to live without anger, fear, worry, and anxiety. Through the practices outlined in the writings of the most famous Stoics such as Epictetus, Seneca, and Marcus Aurelius, we can begin to train our mind to approach troubling situations differently by applying logic and reasoning.

For example, regular practicing of *negative visualization*, where we imagine that our health, our possessions, and loved ones are no longer in our lives, both prepares us mentally for a situation where we are to lose them and also allows us to better appreciate them in the present moment when they're here. Likewise, denying *some* pleasures from ourselves not only allows us to appreciate them more when we do indulge in them but also increases our self-discipline.

Stoicism and Zen Buddhism have a number of parallels: a reduction of internal suffering by improving the interactions that we have with our minds and the world around us, labeling of worldly possessions as negligibly important, and a focus on how to live well now rather than worrying about the future. That's what we're going to look at next.

The Trichotomy of Control

We opened this chapter with our protagonist worrying about the status of important tasks, whether they could have done better in the previous week, and whether or not their team was going to have everything done by Monday.

Epictetus, a famous Stoic, stated that: "some things are up to us and some are not up to us." With reference to our protagonist, the dilemma that is disrupting their tranquility is the fact that they are worrying about things that are *not up to them*: they cannot change the past, nor can they change the outcome of the work that their staff are doing since it's the weekend.

In *A Guide to the Good Life: The Ancient Art of Stoic Joy [Irv09]*, William B. Irvine explains the Stoic approach to control. They reason that we should only focus on things that we can affect. Stoics would argue that our protagonist's problems are fundamentally out of their control, so they shouldn't worry about them. Period.

But, hang on a second. I'm sure that you, the reader, have some objections:

- Surely this is an oversimplification! Everything that I'm going through in my managerial role is much more stressful and taxing than this.

- How can I just stop worrying about outcomes? That's impossible!
- If I do manage to stop worrying about outcomes, then doesn't that make me a terrible manager? What if I cared about nothing in my job at all? Haven't you watched *Office Space*?

We could interpret what Epictetus said as a *dichotomy* of control:

- Things we have control over, such as our desires and goals.
- Things we don't have control over, such as the weather.

However, Irvine recommends splitting things we can't control into two, making it a *trichotomy*:

- Things we have control over, such as our desires and goals.
- Things we truly have no control over, such as the weather.
- Things we have *some, but not complete, control*, such as our desire to win a tennis match.

The latter category is extremely important for you as a manager, since you get your work done by delegating to other people and working through your team. Clearly, you'll worry about whether that work is getting done to an acceptable standard; however, it's being done by other people, and you have to employ the skills that you've learned in this book to get it done to the standard that you wish.

But there are always going to be times that the outcomes that you get from a project or a delegated piece of work are not as you would ideally expect. Perhaps the project was much harder than originally thought. Perhaps it didn't sell as well as the company wanted. Perhaps the task that you fully delegated came back done totally wrong because you wrongly judged the expertise of one of your staff. You could take this on the chin. It could upset you and ruin your weekend. Or even your month. Believe me, I've been there.

This has parallels with somebody taking part in a tennis match: they clearly want to win, else why would they be playing? However, a strong opponent or bad luck can mean that the outcome is not what they wished: they may lose. How can they maintain their inner tranquility?

The answer is that for things that you *do not fully control*, you should set *internal* goals rather than *external* ones. Don't hang your happiness on whether the delegated task comes back exactly as you wanted it. Don't bet all of your emotion on the company getting exactly 15.4% growth this year. You just set yourself up to fail. Those are *external* goals.

Instead, your *internal* goals should be to perform the best that you can, in the situation that you're in, during the time that you're in it. As long as you have done your best as a manager, then you'll always be able to maintain your tranquility. If you ultimately judge yourself on things that you don't have complete control of, you're going to be nothing other than unhappy all of the time.

For example:

- If you're launching a brand-new product and you're extremely worried about whether it's going to be a success (an *external* goal that you cannot control), you shouldn't worry about that. Instead, you and your team should do the best that you can to make it as good as possible (an *internal* goal that you can control).
- If a member of your staff is thinking of leaving and you're concerning yourself with making them not leave (an *external* goal that you cannot control), you shouldn't worry about that. Instead, you should do your best to counteroffer and accommodate them to the best of your ability, but know that you cannot stop them from leaving if they really want to (an *internal* goal that you can control).
- If it's looking like your project is going to be late, you shouldn't worry excessively about it (an *external* goal, as at this point there's nothing that you can do). Instead, you and your team should try your best to complete it as quickly as possible, perhaps by playing with the levers of scope, resources, or time (an *internal* goal that you can control).

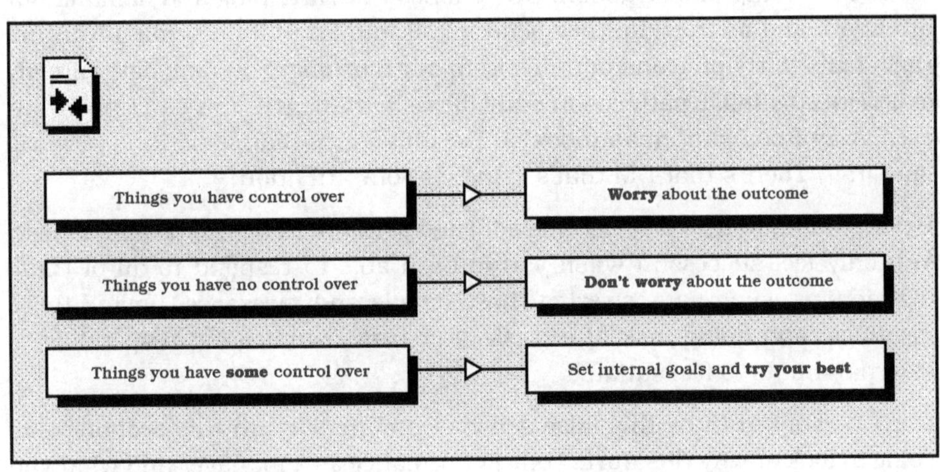

Let go of outcomes that you cannot control. Be accepting of trying your best, and encourage the same behavior in your staff. Unpredictable results are normal. Failure is acceptable. As long as you're trying your best and you're enabling your team to try their best, then you have nothing to worry about. Pat yourself on the back. Epictetus would be proud.

Your Turn: Applied Stoicism

Consider the following situations. How did you act within them, and how did you judge yourself and your team? Did you beat yourself up about it? Do you think that judging yourself by internal goals, rather than external ones, would have made you feel better about these situations? Would doing so have changed the outcome at all?

- A time when you've worked on a project that took way longer than originally expected.
- A time when the company didn't perform well, such as having a bad year for sales.
- A time that you've been made redundant from a role. (If it hasn't happened to you, then imagine it. Cuts were needed, and you got unlucky.)
- Some work that somebody on your team has done that hasn't been up to scratch.

Now, consider what you're working on right now and how you judge yourself for it being successful. Are you on a crash course to ruin your tranquility by setting external rather than internal goals? If so, how can you change them?

Escaping the False Productivity Trap

We live in an age of distraction. Our phones buzz and bing and demand our attention, and as they do, they send us off into an infinite scroll adventure that steals thirty precious minutes from our morning. You may have already experienced similar patterns in your life as a manager. You can't peel your eyes away from your email inbox on the off chance that something new will land in it. There's that DM that's vying for your attention.

Before you know it, you've become addicted to input. You become reactionary. You only feel successful when you've been able to respond to all of those distractions, such as answering those emails and messages, having those meetings, and doing task after task after task. You rely on responding to external inputs to feel satiated.

Don't be sucked in by this false sense of productivity. In this section we're going to look at why this hurts your performance as a manager and what you can do to fix the problem. You need to let go of the behaviors exhibited by one part of your brain to enable the behaviors of another part. Let's find out more about them now.

L-Mode and R-Mode

Your brain is amazing, but it's buggy. We can work out those bugs. In *Pragmatic Thinking and Learning [Hun08]*, the brain is modeled as a dual-CPU, single-master bus computer. Only one of those two CPUs can work at any given time, and they're quite different:

- The first CPU is a slow, linear, traditional Von Neumann architecture. Instructions are processed one after another in order.
- The second CPU is a digital signal processor: it searches and pattern matches asynchronously. It can magically find links between unrelated things, but you have no control over it.

They share the same memory bus, so if one is working, the other isn't. If the first CPU is executing instructions, the second isn't asynchronously searching, and vice versa. The first CPU is called L-mode, where the L stands for *linear*. The second CPU is called R-mode, where the R stands for the *rich* creative and holistic processing it does, as shown in the image on page 252.

You need both of these modes. R-mode helps you innovate, come up with new ideas and connections, and break through walls. L-mode lets you execute the details. Remember that time you magically came up with the solution to a programming problem while you were having a shower? That was R-mode. Then when you got to work, you sat down and wrote each sequential line of code to implement it? That's L-mode. *Thinking, Fast and Slow [Kah12]* similarly defines *System 1* (R-mode) and *System 2* (L-mode) modes of thought.

However, even though we need both of these modes, we quite often don't realize that we need them, nor do we put them to good use. Being a great manager isn't about being an L-mode robot all of the time: answering email after email, checking your messages, answering that DM, checking your emails, ad infinitum. Great managers are also being creative. They're utilizing their R-mode to find new solutions to problems. They're creating space to let their R-mode find new connections: to get that insight about how to make their team better, how to solve that tricky programming problem, or discover some excellent questions to talk through in the coming week's one-to-ones.

Enabling Your R-mode

The thing about R-mode is you can't control what it does. But, you can create the space it needs to do what it does best. To do that you need to carve out time for yourself. To get away from those emails, bleeps and bloops, and distractions. To sit and think and have no agenda. To disengage L-mode and let R-mode run amok.

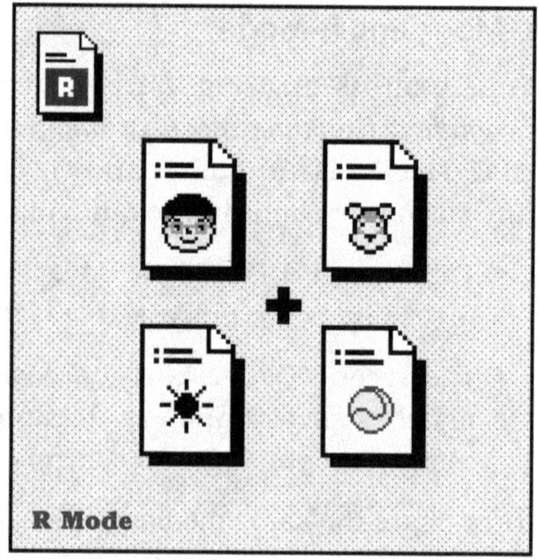

Here's what you should try and do. Try to carve out 10% of your time each week to do absolutely nothing other than let your thoughts emerge. Yes, seriously. You need to block out time in your calendar in which to do nothing. Get away from your desk. No meetings. Maybe even get out of the building and go for a walk. You need to create the conditions of that shower where you suddenly solve the problem, or that drive where you needed to pull over on the highway to write down your brilliant idea. You need to get your R-mode firing.

How you decide to do this is up to you. You know yourself best. Do your best ideas come to you during the beginning of the work day or toward the end? Could Friday afternoon be full of insightful ideas? Pick your times, block them out in your calendar, and do an activity that lets your mind wander. Bring whatever tool you're using as your place to capture information—such as a notebook—outlined in *Manage Yourself First* so that as your mind wanders and up pops an incredible insight, you can note it down and forget about it for now. You can schedule it for L-mode processing later.

Once you reach the end of your blocked-out R-mode period, sit down and go back through the notes that you made. Flesh any of them out if you need to. Commit them to your to-do list, or follow up on any of them that you need to. Perhaps your time away from your desk gave you that insight on how to refactor that particularly gnarly piece of code. Perhaps your brain reminded you that your team is about to tackle a challenge similar to something one of the other teams solved last year and you should go and talk to them. It's funny what pops up when you give yourself the space.

Your Turn: Engage Your R-mode

OK, you've just read about it, so now it's time to do it. Block out some time in your calendar this week to do nothing but let your mind wander. Go and sit on a sofa or go for a stroll. Note anything down that your R-mode surfaces.

When you've done it, how did you find it? What did your brain discover for you when the search results came back? Did you have any insights? Did you make any new mental connections? Will you be doing this more often?

Using Your Capacity Wisely

In addition to purposely blocking out time for R-mode to discover insights, you should be thinking about your capacity more broadly. As you probably already know by now, being a manager can bring all sorts of unexpected twists and turns:

- Sometimes the proverbial hits the fan and you can get interrupted by the production system exploding, with the fallout eating up the rest of your day.
- Perhaps one of your team is having a hard time and needs to confide in you in private, despite the fact you're in the middle of something important.
- Various points of the year trigger extra work such as taking part in the budgeting process and preparing and doing performance reviews.

You need to think about your work capacity and what you commit to. Ideally you should be planning your week and your commitments so that you're operating at a comfortable 85% capacity.

"But," I hear you cry, "isn't my own manager going to think I'm slacking off?" It's a good question. However, as the list above shows, there's always going to be extra work. What it means is that you can take on that extra work willingly, diligently, and complete it to a high standard. If your team needs you, you're available. If your own manager needs some help, you can offer. If one of your staff is getting stuck or is feeling unhappy, then you can easily flex into that spare capacity and be there.

Don't underestimate the power of being able to be there. So many managers operate at capacity that they quickly become weak links in the chain. That ad-hoc request goes unanswered. That member of their staff that needs their extra time doesn't get it. That urgent matter that needs following up gets dropped.

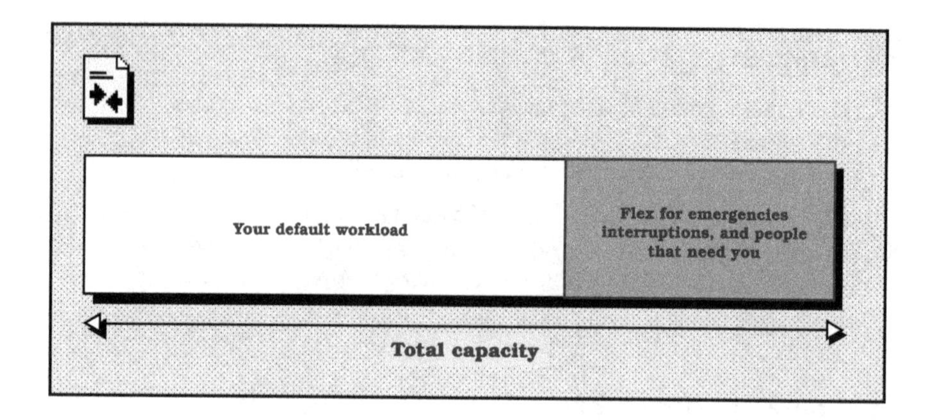

Operating at this capacity requires some changes:

- Perhaps you could *block out even more time* in your calendar for thinking and reacting. Try a thirty-minute slot at the beginning and the end of each day.
- *Purposefully take on 85% of the work* that you think that you can get done, especially when it comes to planning sprints. You can look back at your to-do list and calendar over a month period and categorize your activities. You can then scrutinize them to see whether or not they were a good use of your time. Then cut out anything not worthwhile going forward.
- You'll have to learn to *deal with the guilt* of taking time for yourself. It's real and it will be uncomfortable. However, with time, the ability to react to unexpected events quickly, and the additional insights that your R-mode brings to you, will make you wonder why you didn't do this sooner.

So, you're being told to work a bit less hard. But actually, you're being wiser with your time and letting yourself work more effectively. Trust me. It'll make you a better manager. *Let go of needing to fill every minute of your week.*

Stop Checking Those Notifications!

If it isn't clear by now, the magic begins to happen when you free up some time, allow your R-mode to engage, and have more space to do your job in a holistic and creative way. However, we opened this section talking about the distractions of everyday working life. The push notifications, the emails, the buzzes, bleeps, and bloops.

They're addictive. Many pieces of software are designed to keep you coming back for more, triggering those little dopamine hits as you clear off your notifications, answer your messages, and achieve inbox zen. In addition to managing yourself and the way that you use these tools, you need to *manage how often you use them* in the first place. It's likely you've spent most of your life hooked to the drip feed, so it's time to disconnect.

Limit the amount of times that you open your email inbox to just a few mindful occasions every day. Do the same with your chat software, your instant messaging clients, and whatever else you use that distracts your attention. See what times of the day work best for you to batch your communication. Close them at all other times. Perhaps you could answer your messages once in the morning with a coffee, once before lunch, and then have a clear out at the end of the day before you stop work. See what's best for you. But don't keep them open. The eyes wander, the mind wanders, and the dopamine hits of keeping on top of the flow of conversations will keep you from really making an impact in your job.

Let go of compulsive checking. Resist, resist, resist.

Your Turn: Don't Check Those Messages

Here's a tough one. Tomorrow, see how long you can go at work without checking any emails or messages at all. Answer your important messages first thing and then leave the rest and close any tabs and applications. Yep, that's right. Don't look.

How long did you manage to go without opening them again? How many times did you notice yourself wanting to check? What did that urge feel like? Was it emotional? Did you feel like you were missing out? When you eventually opened your emails or messages, had anything actually happened of any criticality?

What You Do Outside of Work Matters

And now to the big one: *letting go of work itself.* As much as those people on Twitter with tens of thousands of followers and verified accounts are telling you to keep hustling, to be on top of your game, and perhaps even wake up at 4:00 a.m. in order to be as successful as the world's richest CEOs, don't be fooled: they're just trying to sell you something. You need to look after yourself.

In this chapter we've already covered a myriad of ways in which letting go can benefit you: from delegation, to not worrying about outcomes that you can't control, to better managing your capacity, and also by creating the conditions for your R-mode brain to discover new connections, patterns, and innovations.

However, all of these are just techniques for you while you're *at* work. You also need to make sure that you're looking after yourself after work. We're only on this planet for a certain amount of time, and that time is fleeting. Work is a part of life. But many other important parts get neglected at the expense of work: through working long hours, to being unable to stop those thoughts about that annoying conversation spinning around your head all evening, through to—in the worst case—the recovery time needed after burning out. You need to look after yourself.

Statistics from 2017 [Mur18] show that about 13% of the global population (around 971 million people) suffer from some kind of mental disorder. Just under 300 million suffer from anxiety, around 160 million suffer a major depressive disorder, and 100 million experience a milder form of depression called dysthymia.[1]

In addition to these medical conditions, *burn-out* was included in the World Health Organization's International Classification of Diseases in 2019 as an occupational phenomenon.[2] They describe burn-out as a "syndrome conceptualized as resulting from chronic workplace stress that has not been successfully managed."

Burn-out manifests as:

- Feelings of energy depletion or exhaustion.
- Increased mental distance from one's job, or feelings of negativism or cynicism related to one's job.
- Reduced professional efficacy.

Clearly, we need to take this seriously so we reduce our chances of burn-out and mental health issues. The good news is that the way in which we need to look after ourselves is simple to understand. The bad news is that we have an industry, culture, and society that is making it harder to do so.

If you do think that you are suffering from a mental illness such as depression, anxiety, or both, then you should talk to somebody about your feelings. It isn't a sign of weakness. Dealing with existing mental health issues is beyond the scope of this book, but you should talk to your partner, friends, family, your manager, your HR department, or seek advice from a mental health practitioner or therapist. More people do this than you think. I had a particularly difficult couple of years in which I was supported by people in all of these roles. Those that didn't know I was going through it had no idea.

---

1. http://www.healthdata.org/
2. https://www.who.int/mental_health/evidence/burn-out/en/

You can do many things for yourself that can positively contribute to your mental well-being. Those are straightforward suggestions that we definitely can cover in this book.

We're going to look at:

- *Getting enough sleep.* The biggest predictor of whether or not you're going to have a good day is the amount of sleep you're getting. We'll look at why and at some techniques that you can use to improve your sleeping habits.
- *Getting regular exercise.* Humans haven't evolved to sit at a desk all day. We need to move around as well. We'll look at some simple ways to integrate some movement into your day.
- *Being present in the moment.* We continually live in the past and the future but increasingly find it hard to give our undivided attention to this moment, right now. We'll see how you can train this muscle.

You've got to manage yourself to manage others well. Let's start by looking at sleep.

The Importance of Sleep

If there's one thing that you take away from this chapter, it's that you should be, as much as possible, getting eight hours of sleep every night. A good night's sleep is the most reliable way of ensuring that you approach the next work day in a calm, thoughtful, and cognitively balanced manner.

Matthew Walker's *Why We Sleep [Wal17]* highlights the critical importance of sleep in humans. Even though studies have shown that insufficient sleep results in reduced concentration and cognitive performance and an increased risk of neurological and psychiatric conditions such as Alzheimer's, strokes, chronic pain, and mental health issues, we routinely do not get enough. In fact, our habits are getting worse. Were you watching videos on your phone last night in bed before you went to sleep?

A startling correlation has shown that the shorter your sleep, the shorter your lifespan. Getting a proper amount of sleep carries the following benefits:

- Improved motor skills, notably in speed and accuracy.
- Increased time to physical exhaustion and increased aerobic output.
- More stable emotional states.
- Improved immune system function.
- Improved R-mode activity and, as a result, increased creativity.

You're in an industry where it's critical to be able to continually learn new skills. Professor Walker explains that it isn't just practice that makes perfect: practice and a *good night's sleep* makes perfect. Subjects who have just learned a skill have been measured as being 20–30% better at the skills after sleeping when compared to measurements after the learning session the day before. As he puts it: sleep is the best legal performance-enhancing drug that we have.

There's no such thing as catching up on sleep either. Studies have shown that humans can't counteract a week of poor sleep by sleeping more on the weekend. Sleep is a one-time offer. You lose it, you lose out. (New parents, caregivers, or those with sick children, loved ones, or pets: I'm sorry. I know you can only do your best to get enough sleep in times of need. In these situations you are sacrificing your own sleep for the benefit of others, which is clearly the greater good.)

So what are some practical things that you can do to improve your sleep?

- *Limit your caffeine consumption.* Caffeine blocks the receptors in your brain that tell you that you need to sleep, pushing you to stay alert when your body really needs you to sleep. If you can't go cold turkey with coffee, then try and only drink it in the morning.

- *Limit your alcohol consumption.* Although the "nightcap"—a late-night alcoholic drink to help you sleep—may have entered societal lore, it actually makes your sleep worse. Although it may help you lose *consciousness*, it doesn't allow you to have good *sleep*. Nights after drinking alcohol are fragmented by frequent periods of waking up that are so small that they're not committed to memory. You'll wake up feeling tired and unrefreshed.

- *Keep your gadgets away from the bedroom.* Using phones, laptops, and tablets in bed will extend the amount of time that your body takes to release melatonin by up to 50%. You'll be on a neighboring time zone before you know it.

- *Have a cool bedroom.* A cooler core body temperature helps you sleep well. Keep your bedroom at around 65 degrees Fahrenheit for optimal sleep conditions. A hot bath doesn't make you sleepy because you're hot afterward; it's because the heat moves your blood closer to your skin, which means you *cool down quicker*, ready for sleep.

- *Don't exercise too close to bed.* Exercise is great for helping you sleep better, but not when you do it only a few hours before bed. It'll keep you awake.

- *Try to sleep and wake at the same times each day.* Have a sleep routine and a sleep schedule. Set yourself up for success by forming a habit.

Slowly wind down your day, put your devices aside, stop checking messages and notifications, and go and get comfortable.

If you could improve your performance tomorrow by doing one thing, wouldn't you do it? Go and get a good night's sleep. *Let go of wanting to fill every waking hour with productivity.* You steal from the future to do more in the present. More sleep will make you better in life and work, and it may help you live longer too.

Your Turn: Audit Your Sleeping Habits

How healthy are your sleeping habits right now? Go back after the last week and think about how well you've been sleeping. What did you find? Are you sleeping enough, or too little? Is that preventable or not?

For the coming week, make a conscious effort to follow the advice we covered and improve your sleep patterns. Instill a sleeping habit. Moderate your alcohol and caffeine consumption. Stop using those gadgets before bed. Wind down your evenings. You won't regret it.

Moving Your Body

Humans have not evolved to spend their entire day sitting. In fact, studies *have proven [Kut13]* that it isn't good for us at all. However, as a society, we seem to love sitting down. We get out of bed, sit while commuting, sit at our desks all day, sit down on the commute home, and then spend the evening sitting on the sofa. This is bad for our physical and mental health, which not only affects our performance at work—it affects the entirety of our lives.

But the good news is that we don't need to drastically change our lives to get the exercise that we need. You can hold off on ordering those sweatbands and spandex leotards for now. You don't even need a gym membership. You can begin with small steps and work toward larger goals iteratively.

- *Consider walking or cycling to work.* Not only is it beneficial to the environment, you'll get a load of beneficial exercise done without needing to specifically allocate the time.

- *Move for five minutes every hour.* Just get up from your desk and go for a little walk. Do a lap around the office, or if you're working at home, be creative.

- *Turn sitting meetings into walking meetings.* Why not do your one-to-ones while taking a walk? The act of walking *has been shown [OS14]* to increase creative thinking, is meaningful exercise, and also gives you more to look at than the meeting room wall.

- *Slowly build toward bigger things.* As you do more walking, or more cycling, you'll begin to feel the positive benefits. You'll sleep better, you'll feel more relaxed, and you'll benefit from those endorphins. Think about ways in which you can keep adding more movement into your life. Perhaps you could give running a go. Maybe you'd like to have a go at lifting weights or joining an exercise class. Either way, it's all good. Just keep moving and doing what you enjoy.

Like sleeping, exercise is nature's way of helping you get stronger, and it also positively affects your mental health by releasing endorphins. Get on board. *Let go of your chair, no matter how comfortable it is.* Consciously build in more movement into your day. Get your blood pumping. You'll feel sharper and happier.

Your Turn: Get Moving

If you aren't already exercising regularly, introduce a small amount of activity into your day. You could change your commute to work, or if you'd like to start even smaller, just go for a short walk in the evening or at lunchtime.

The key is building a habit where you want to get moving without even having to push yourself to do so. Start small, and make incremental improvements. Give yourself some simple goals, such as beginning by doing something that adds up to thirty minutes of movement each day. You'll be hooked in no time.

The Only Moment That Matters Is Now

Our Stoic friend Epictetus (remember him?) stated that some things are up to us and some things are not up to us. Stoics, partly for rational reasons and partly as a product of their time, believed in fate. From the ancient Greek myths, fate was represented by the Three Fates: Clotho, who spun the thread of life, Lachesis, who measured it, and Atropos, who cut it. Ancient Greeks believed that our destinies were already mapped out for us, thus reducing the need to worry about all of the eventualities of life. Modern religions have similar parallels, where there's a belief that one's inevitable life trajectory is in the hands of one or more deities.

One interpretation of this fatalism advocates for being fatalistic with respect to the past—whatever has happened has happened and cannot be changed—and also to the present moment. It may be possible to perform actions that affect the future, but the present moment is by definition already happening, so we're merely an observer.

This has a relation to how Zen Buddhism treats the present. In fact, Buddhists, and more increasingly the secular West, have been learning about the concept of *mindfulness*. Jon Kabat-Zinn is considered to be responsible for the popularization of Western mindfulness throughout the past few decades; not through marketing, but through science. His seminal book, *Full Catastrophe Living [Kab13]*, outlines the work done at the Mindfulness-Based Stress Reduction Clinic at the University of Massachusetts Medical School. The eight-week course was a supplement for patients with chronic pain, terminal illnesses, and disabilities. It's meditation and mindfulness with the Buddhism taken out of it: no religion required.

The main components of this course are simple mindful breathing exercises, meditation, yoga, and body scan techniques: all things that you can do in the peace of your own home for free. *Two decades of research have shown [Sre05]* that those that regularly practice mindfulness techniques have:

- Lasting decreases in physical and psychological symptoms.
- An increased ability to relax.
- Reduced pain levels and improved ability to cope with lasting pain.
- Greater energy and enthusiasm for life.
- Improved self-esteem.
- An ability to cope more effectively with stressful situations.

Not bad for a few simple exercises. A wealth of information is out there for you to learn about applying these mindfulness techniques. A good beginner's guide is Jon Kabat-Zinn's *Wherever You Go, There You Are [Kab16]*. To get started though, why don't we quickly do something together?

Letting Go…of This Chapter

There we are: you've discovered a myriad of reasons and techniques for letting go a little more. The irony is, letting go will both improve your effectiveness as a manager and also improve your physical and mental well-being.

In this chapter, we've looked at the following:

- *How to let go of tasks.* Being an effective manager is all about letting go via delegation to improve and empower your teams. However, we also looked at how even though you may be delegating well, you might still be worrying. We took some inspiration from the Stoics to reframe how we think about things that we control, things that we do not control, and

Your Turn: Mindful Breathing

This exercise is intended to help you relax by mindfully bringing your attention to your breath. The idea is that your mind calms as you focus your attention to what you are feeling. Set a timer to help you focus on the exercise, or alternatively download a meditation timer app for your phone.

- Lie down on your back or sit erect in a comfortable warm place. Close your eyes (but read the following instructions first!).
- Turn your attention to your breath, feeling it move in and out of your body.
- Focus on the effect that your breath has on your belly: feel it rise, and fall. Stay here for a minute.
- If random mental chatter and thoughts are coming into your mind, don't worry, that's normal. Observe them and then bring your attention back to your breath.
- Now focus on the breath as it passes in and out of your nostrils. Stay here for a minute.
- Gently open your eyes.

How did you find that quick meditation? Were you calm, or was your mind extremely busy? Try and find a few minutes each day to practice this. If you kept getting distracted, then do know that it's normal. It's an unfortunate bug in your brain. Left unattended with no input, it creates its own stream of chatter. It takes a lot of time to get used to not clinging on to distractions generated by your brain as they are produced. As Jon Kabat-Zinn *says [Kab16]*: "try it for a few years and see what happens."

things that we have some control over. We learned how to set internal rather than external goals to not set ourselves up for failure.

- *How to let go of your preconceptions for effective work.* We explored the L-mode and R-mode of your brain, and how not carving out time for yourself at work may be seriously hampering your creativity. We also saw how setting aside this time can make you a more effective manager by giving you flex to react to emergencies, interruptions, and people that need you.
- *How to let go of work after work.* We learned some simple techniques for improving your body and mind. You saw the importance of sleep, exercise, and mindfulness as tools to make you a better manager and human being.

Now that you're recharged and raring to go, we're going to focus back on your job in the next chapter. Notably, we're going to begin to consider how you can create alignment with information sharing and best practices outside of your team, even if all of those people are not necessarily working together or within your control. It's time to learn about *good housekeeping* in your department.

Tidy home, tidy mind.

> *Anon.*

CHAPTER 14

Good Housekeeping

Production issues again. You and your team are sitting together looking through the logs and scratching your heads.

"That just doesn't make any sense, though," says Ben. "Why would the quorum keep trying to connect to the instance that crashed?"

"No idea," you reply. "Isn't it meant to detect that it's gone down, then another instance gets elected as leader?"

"Hang on, didn't this happen before?" says Tara. "I seem to remember us debugging a similar issue a few months back. What did we do to fix it then?"

You can't remember. You search back through old commits to the codebase to try and find the one that was meant to have fixed this problem last time, assuming it did actually happen last time, and also assuming that you did actually fix it. No dice. You can't see anything.

"Can't find a relevant commit, unless it's named something silly," you say.

You notice Emma walking past—she's a principal engineer and no doubt would have some good insight into this problem.

"Emma, have you got a second?"

"Sure, what's up?"

"We're having production issues with our storage layer again. Some of the instances have died, and we've brought them back up, but the existing instances won't stop connecting to the old one, and our apps are stuck as a result."

Emma wheels a chair over and has a look at the logs with you.

"OK, can you show me the configuration that you're launching this with?"

"Sure," you say, bringing up the production configuration to the application. Emma leans in and studies the values.

"Where's the connection timeout default defined?" she asks.

"There's meant to be a default?" you reply.

"Yeah. Otherwise it'll retry dead instances forever and not try to seek a new leader. I think that's your problem. Put that in and restart it and see what happens."

You do that, and the cluster elects a new leader and springs back to life. Your applications go green again.

"I'm surprised you're not using the defaults we're all using, since that's inherited from there," says Emma as she stands up from her chair.

"There's defaults?" replies Tara.

"Yeah, there's a parent config for applications talking to that storage. I'll send you the link to it."

"Do you reckon this is what caused all of the outages that we had last month?" you ask.

"Probably, it seems likely," says Emma, walking back to her desk. "I'll send you the location of the parent config now."

The members of your team look at each other, clearly feeling a bit stupid.

"Did we really lose all of that time last month because we didn't have some config values set?" asks Ben.

"Seems like it. We spent days thinking it was our code," you reply. "Why didn't we know that config existed? Maybe even after the third time it happened?"

Does this situation sound familiar? We've all been there. Some missed communication, bad documentation, or crossed wires causing all manner of problems. This happens at all levels: between individuals, within teams, and as a whole department and company. Without a disciplined approach to improvement, communication, and information sharing, departments can begin to fragment. Teams can feel like they are speaking different languages, even though they're sitting next to each other in the office.

The good news is that there are a number of techniques that you can use as a manager to keep your house tidy. You don't have to wait for your CTO to make declarations on how to solve these problems from the top down. Instead, you can begin to implement them in your team for an immediate benefit. You can then help spread this bottom-up approach by sharing what you're doing with others in your department and bringing them on board. Your spotless house can make the neighborhood better too.

This chapter is about forming groups and instilling simple processes so that you embrace collective learning, make fewer repeated mistakes, and interact with others that you may have never met before. This chapter is about good housekeeping in your team and in your department.

Here's what we're going to look at:

- Why communication is tricky, especially in large departments, and how that has *a measurable effect on the software that you build.*
- How you can cross team boundaries to unite people of similar skill sets and interests with *guilds.*
- How encouraging a culture of *technical talks* can unite your team and your department.
- How *problems can be opportunities* for department-wide improvement, both in your software and in your processes.
- Some *tools for solving common problems* such as understanding the context around why software is designed in a particular way, whether a team is improving over time, and working out who is responsible for what.

If you're looking for ways to increase your influence in your department, then this chapter might just be the one for you. You'll find many practices you can start in your team that can pick up momentum elsewhere, and you can be the driving force. Are you ready? Let's go.

Communication Dictates Software Design

If you get communication right within a department, there's a wealth of benefit to be had. Notably:

- *Individuals are aligned* with the collective purpose and vision supporting what is being built, and each individual understands the part that they and their team are playing toward achieving it.

- *The department learns collectively* as progress and new discoveries are shared, individuals normalize on best practice, and mistakes generate better ways of approaching those situations in the future.
- *Individuals feel engaged and motivated*, since they have maximal opportunities for learning from others and also broadcasting what they have learned.

However, doing it right is tough. We touched upon the difficulty of communication earlier in *Projects Are Hard*: we noted that the number of communication channels far outweigh the number of staff as an organization grows. But the *number* of communication channels are only part of the problem. In fact, getting communication right among any group of people—whether that's your team, your department, or your company—begins to sound a whole lot like computer networking: more specifically, *routing*.

The routing of messages between nodes in a network is performed by a particular delivery scheme. How we communicate as individuals at work has clear parallels. Examples of delivery schemes are:

- *Unicast*, which delivers a message to a single specific node. Think of this as sending a DM or an email or having a private conversation.
- *Broadcast*, which delivers a message to all nodes in the network. This is similar to an all-hands meeting or an email to the staff-wide mailing list.
- *Multicast*, which delivers a message to a group of nodes that have expressed interest in receiving the message. This is like a skill-set–specific public chat room such as #backend or #datascience.
- *Anycast*, which delivers a message to any node out of a group, such as asking someone walking past if they can sense check how you're trying to debug a problem.

Getting communication right involves instilling some principles, techniques, and tools for getting the right groups of nodes receiving information using the right delivery scheme. In addition to the benefits of alignment, learning, and employee engagement that we mentioned previously, the way in which communication occurs can even have an effect on the design of your software.

In 1967, the computer scientist Melvin Conway stated that "organizations that design systems are constrained to produce designs which are copies of the communication structures of these organizations." This is commonly known as Conway's Law.[1] One of a number of fun variations on this statement

1. https://en.wikipedia.org/wiki/Conway%27s_law

is Eric S. Raymond's: "If you have four groups working on a compiler, you'll get a 4-pass compiler."

Although it is conjecture, Conway's Law sheds light on some key corollaries. As a new manager, it's unlikely that you've been able to design the department from the ground up. You've got your team, and you're learning how to work with them and also how to work with the wider department. You may be looking out across a department filled with different types of teams, such as:

- *Skill-set teams*, which work well at concentrating expertise and ensuring best practice with a given skill set or technology, since all of the experts are working together and communicating regularly.
- *Cross-functional teams*, which work well toward a common goal—such as a completely new feature in your application—since the team can handle the whole life cycle of the project. Cross-functional teams are common in product-centric companies, since shipping a new feature often needs input from design, product management, UX, back end, front end, and QA.

Skill-set teams struggle when the company needs to build software that bridges multiple skill sets—which is most of the time—as development effort involves coordinating multiple teams and interfaces. They can easily become silos that embody a physical manifestation of the software stack.

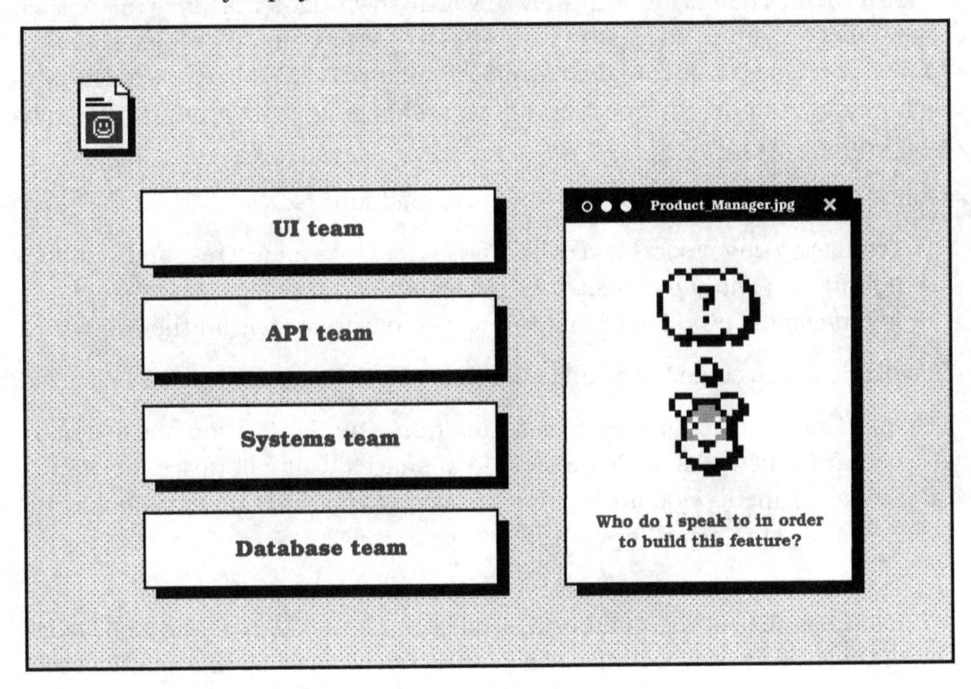

Our industry, especially with the influence of agile development, recommends cross-functional teams. But there's no silver bullet—they can still be subject to the effects of Conway's Law. How many times have you experienced two feature teams building the same thing twice? What about when they build two variations of the same piece of functionality that then creates technical debt in the main codebase? It happens. Cross-functional feature teams without effective communication are *just another type of silo*. You need to be careful.

If it's the case you get silos with skill-set teams and silos with cross-functional teams, is there really nothing that you can do? Well, there's a neat solution to this problem: creating *guilds*.

Breaking Silos with Guilds

Guilds are a straightforward yet powerful technique for preventing silos. A guild is a group of people who work on different teams but share a common interest or skill set. The neat thing about guilds is that since they're just a group of people with a common goal, and since it's so easy to talk via chat and video call, you too can help form guilds in your company: no top-down approach required.

You may be wondering why we're covering guilds if they don't need a manager to form them. The reason that they're worth covering is that they're another useful tool for you to suggest to your staff to solve common problems, which in turn increases their confidence and autonomy. You're also well-positioned as a manager to help connect others as guilds get off the ground. You may also want to get involved yourself!

Guilds concern themselves with the following matters:

- *Discussion and progression of best practice* across an interest or discipline.
- *Information sharing* across multiple teams.
- *Improving the visibility* of that interest or discipline within the company.

For example, you could have guilds that concern themselves with the following:

- How the department uses *JavaScript*, from standardization of use of particular frameworks, code ownership of shared utility libraries, to owning the department-wide linting configuration to ensure that code looks and feels the same regardless of the author.

- Visibility and promotion of *security* throughout the development process. Each member of the guild brings back the best practice and knowledge to their own team and raises any concerns from the team up to the guild to discuss.

- Unity of *design and UX* practices across different teams and products.
- Championing of *diversity and inclusivity* across teams, culture, and the hiring process.

Guilds don't have to be all about code and frameworks. In theory they can be about anything that contributes to the company and its culture. One of the most widely known implementations of guilds is at Spotify,[2] where guilds are only a small part of how they subdivide agile teams. Since Spotify is a large and forward-thinking company with regards to team structure and agile methodologies, they have the following named groups of people:

- *Squads*, which are cross-functional development teams.
- *Tribes*, which are collections of squads that work in a related area, such as infrastructure.
- *Chapters*, which are groups of individuals of a common skill set *within tribes*.
- *Guilds*, which are groups of individuals across the whole department with a common interest.

It's easier to collapse the latter two into just "guilds" for the purposes of this book and for your initial implementation. The Spotify implementation of this structure also has implications for line management, which aren't necessarily of interest to us right now, but you can read their white paper in the footnotes if you're interested. For now, let's just stick to a guild being a group of people that share a similar skill set or interest.

Helping to form guilds may be a good choice for you if you hear the following common concerns:

- That teams are reinventing the wheel because they were unaware that there was code to reuse elsewhere.
- That technology choices are beginning to diverge because it's easier for teams to go their own way rather than learning about what's going on elsewhere in the department.
- That there are unexpected surprises (with design, compatibility, speed, and so on) when teams commit their work into the main codebase.
- That engineers are feeling disconnected from what other teams are working on.

---

2. https://blog.crisp.se/wp-content/uploads/2012/11/SpotifyScaling.pdf

Guilds can help you out greatly here, giving those that participate a chance to share information, raise issues, and have others help work through individual and collective problems.

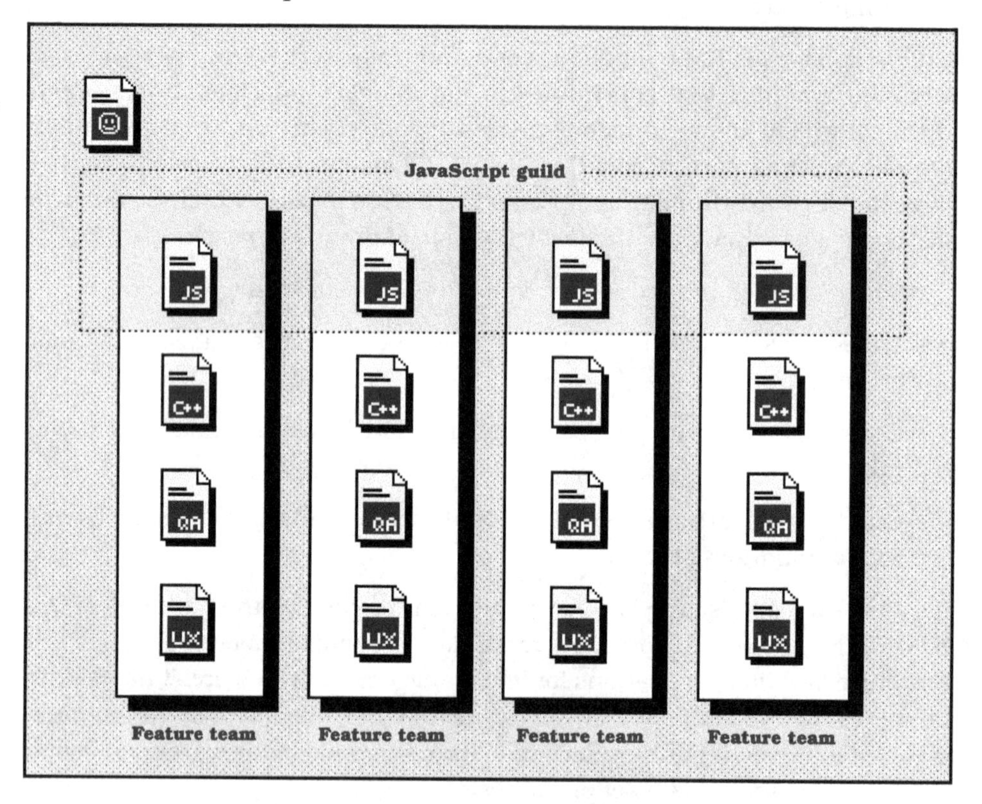

If you're thinking of starting a guild for an interest or skill set then it's as simple as having a conversation with others who share that interest or skill set and beginning to meet up. Here's a handy list of things that you can do that can get a guild off the ground. Remember that it doesn't have to be *you* that gets a guild started. This is simply a tool that you now know to encourage anyone within your team or even your whole department to try out.

- *Announce the guild's intentions.* Simply sending out an email or chat message is a good way of getting started. Outline what the guild is for, highlight some of the reasons that it would be beneficial, and solicit interest from others in the department. Guilds can operate across a spectrum, spanning from *sharing and discussion* to *action and delivery.* Make it clear where this particular guild sits.

- *Choose the members.* From those interested, try and get a good balance of people so that as many teams are represented as possible so that the core group isn't *too* large (it helps to have the core guild function represented by a smaller group, but anyone else can contribute) and so that potential members will have the time to contribute properly. It's natural to think that a JavaScript guild needs all of the most senior engineers to be part of it, but less experienced engineers can learn a lot by participating and also can have more motivation and time to contribute.

- *Choose a leader.* The leader doesn't have any explicit power over anyone else, but someone needs to be responsible for setting up and running the meetings, coordinating the discussion, and keeping documentation up-to-date.

- *Choose a time to meet.* Depending on the people in the guild, it may be that most discussion happens asynchronously. However, having touchpoints where the guild gets together to discuss pressing issues, even if it's just once a month, is invaluable for creating rapport between the members and also giving the guild momentum.

- *Create the communication channels.* Again, your mileage may vary. Perhaps a public chat channel would suffice so that anyone can participate. Maybe a mailing list would be better suited to the company. Optimize for reaching the largest number of people with your communications, since that increases the output and impact of the guild.

- *Decide how to document the guild's work.* Think about where you'd like to store any documentation related to the guild, such as meeting minutes, information about the guild and how to contribute, and the outcomes of any decisions that have been made.

Guilds may not last forever. Sometimes a guild has served its purpose and may need disbanding or merging into other guilds. Alternatively, guilds may grow and need partitioning into subsets with more specialized focuses.

Going back to our networking analogy, guilds are a good example of *multicast* communication patterns in your department. As you participate in guilds, think about their output. How can you spread the knowledge and progress of the guild further? Perhaps you could have a monthly *broadcast* email that keeps the rest of the department up-to-date with what you've been working on. Another good broadcast strategy that you can start from the bottom up is getting technical talks going. We'll look at that next.

Example: Your First Guild Meeting

Here are some ideas for your first guild meeting.

- Have the guild leader create a shared agenda document to capture everyone's thoughts as to what should be discussed the first time that you meet.
- If any of the previous questions about how often to meet, what the best communication channels are, and so on, are undecided, then these are great first topics of discussion.
- Spend time thinking about the lay of the land. If this is the QA guild, what can each member say about how QA is done in their team? Are processes and frameworks the same or different? Is this good or bad? What does a collaborative future look like?
- If this is a technology-related guild, could the guild keep a technology radar—of which tools to adopt, trial, assess, or hold off on—up-to-date with the technologies that they suggest the department should and shouldn't use in that area?
- If this is a cultural guild such as one encompassing inclusivity and diversity, then what do people feel like the pressing issues are to focus on? We'll pick up these themes in *The Modern Workplace*.

Sometimes it can be useful not to overplan the first meeting. You'll already have a group of engaged people with a shared interest in the room together, so maybe just get together and see what happens!

Encouraging a Culture of Talks

Human beings love to share with each other. We enjoy uploading our photographs to social media so that others can see them and comment on them. Young toddlers consistently point at objects and call out what they think it is to get your attention. We also enjoy movies, books, and theater, which are a more formal way of sharing a story or information. I'm sharing this moment with you right now, even if I don't know who you are.

In our industry, hundreds of conferences every year promise days filled with stimulating talks about the latest and greatest technology and practices. People come from all over the world to learn what the speakers have to say. But have you ever considered the collective knowledge of everyone that you work with? What if there was a way to create your own technical talks track so that you and many others can benefit from the knowledge, views, and storytelling capability of others? Well, you can do just that.

The annoying thing about becoming a good speaker is that it requires a lot of practice. Many people simply do not have enough opportunities to get up in front of a group of people and present, which means that they may pass up opportunities that could be beneficial for themselves and their careers in the future, such as submitting a talk to a conference, or being able to confidently present to the board, or perhaps taking a slot at the company all-hands meeting.

This why people usually find giving talks scary. They've simply not had enough practice to become comfortable and confident. This is where you can come in as a manager. You can begin to create the environment where it's safe for people to begin to give talks to each other. You can start in your team. If it goes well, there's nothing stopping you from creating—or helping co-create—a regular talk slot that brings your department together to sit, listen, expand their knowledge, and nurture the collective consciousness.

Team Lightning Talks

Let's start small with a couple of ways in which you can encourage talks from within your team. Giving long and detailed talks requires a significant investment of time to incubate the ideas, distill them into a narrative and practice until they're perfect. To counter this, you can start by introducing a regular lightning talk slot that those on your team can rotate through.

If you haven't previously heard of lightning talks, they're very short presentations lasting, typically, five minutes. The goal is to have the speaker cover a topic in a quick, concise, and efficient manner. They're fun to watch and fun to give.

Pick a regular slot in which you can start doing talks within your team. Perhaps a good point is at the end of every sprint or iteration of your project, picking the speaker at the start of that time period. You can also take some of the stress out of the equation for the speaker by not requiring them to make their presentation standing in front of the team in a large room with an AV setup. You could do it informally, huddled round a laptop, or perhaps you could all do it separately via video chat if you have a globally distributed team.

To keep the lightning talks entertaining and useful, here are some tips:

- *Allow preparation time.* If you're asking your team to participate, then try to ensure that you give them the time to do the preparation work. A five-minute talk will probably take thirty minutes to an hour to prepare. One method is to put a ticket for the preparation of the talk into your ticket-tracking system so that they can assign themselves to it and work on it in the open, rather than have it competing with their other work. These talks benefit the team!

- *Keep slides simple.* A common mistake that speakers make is to overburden slides with content. People in the audience of the talk will be able to either read the slides or listen, but not both at the same time. Since lightning talks are about concise and clear delivery, a recommendation is to keep the absolute bare minimum information on each slide, or even not use slides at all. If you're looking for inspiration here, then check out the Takahashi method,[3] which involves just putting a couple of words in each slide. You might also look at the PechaKucha method,[4] which uses images and spoken word instead of text.
- *Ensure sufficient practice.* Encourage your speakers to run through their talk multiple times before giving it. They could book a meeting room and run through it multiple times to an imaginary audience to get confident. You could even sit in and give them some feedback if it's helpful for them.
- *Have a system of feedback for the speaker.* Since lightning talks are a great way for everyone to gain valuable presentation skills, incorporate a way for those in the team to give their feedback. One option here is to use a survey form, sent around after the lightning talk, to capture one piece of praise and one piece of constructive criticism for the speaker. That way the speaker can improve their technique, and the team can become more comfortable at delivering feedback. Whether the survey is anonymous or not is up to you. If you staff aren't practiced in giving candid opinions then you can start with anonymous feedback to make them feel comfortable. You should build toward them being able to give it directly, though.

Department Talks

If your lightning talks gain traction, then you could consider expanding talk slots out into your department, soliciting input from anyone in any team. You can do this in a number of ways, depending on the appetite of those in your department to engage. A good place to start is to ask whether anyone in the department would like to take ownership of organizing the talks. This could be you to begin with, but it's also an opportunity for someone looking for additional ways to contribute.

3. https://en.wikipedia.org/wiki/Takahashi_method
4. https://en.wikipedia.org/wiki/PechaKucha

Your Turn: Start Doing Lightning Talks

It's time to get lightning talks going within your team.

- Pick a time and cadence that works for you and your team.
- Communicate why lightning talks are beneficial and start drawing up a rota.
- Lead from the front by taking the first slot yourself!
- Prepare your talk. Perhaps a good first talk could be on how to give good lightning talks! It's relevant, educational, and allows you to lead by example.
- Do it! Make it fun, informative, and energetic. Don't worry about formalities. Try to enjoy the process.
- Get feedback. How did it go? Did you do better or worse than you expected? Did you have any habits as a speaker that others picked up on?
- Pay it forward by helping set up the next speaker. Perhaps they could give a short talk on their favorite feature in their programming language, on a particularly interesting problem they solved recently, or even on their role model or favorite project they've ever done. The content doesn't matter as long as it's informative and entertaining.

Once everyone in the team has had a turn doing a lightning talk, ask their opinions. Are they enjoying them? Would they like to continue? Do they think that lightning talks should be opened up to a wider audience in the department and perhaps also have participants from outside the team?

Here's what they'll need to consider:

- *Booking a regular time slot* that the talks will take place (for example, biweekly).
- *Soliciting ideas for talks* and keeping the talks agenda up-to-date.
- *Advertising upcoming talks* to ensure that people attend.
- *Ensuring that the AV setup works* for both physically present and distributed team members. (Most videoconferencing software works fine.) Depending on the size of your company, this may involve coordinating with IT or whoever handles booking the presentation spaces in your office.
- *Recording the presentations* so that they can be shared later for those that were unable to attend.

A good length for internal technical talks is about twenty to thirty minutes. As long as talks are concise—in a similar manner to lightning talks—then a lot of material can be covered within that time. The main benefit of these talks is to raise the visibility of different people, ideas, and projects within the department. A lot of the valuable connections and conversation comes afterward.

Here's a bunch of different ideas for talks that you can offer as suggestions:

- *Show and tells* for projects that are currently underway or have recently been completed.
- *Deep dives* into the inner workings of products and architecture from those that work on it.
- *Success stories* of how individuals or teams solved particular challenging problems.
- *Horror stories* that highlight what happens when everything goes wrong.
- *Introductions to new technologies* for those that are learning something new either at work or for fun.
- *A collection of lightning talks* from multiple participants, batched together in one session.
- *Cultural and well-being elements*, such as dealing with stress, avoiding burnout, becoming a new parent, and so on.
- *External speakers* can even come in and present. If anyone in your department knows anyone interesting, then an external perspective can be extremely valuable. How is another company solving a particular problem? What can you learn from them?

Interesting topics are limitless, and department talks can become a fixture that staff really look forward to. It's an opportunity to gather, take a break from work, and learn. It's also a fantastic way for speakers to practice their skills in front of a friendly, but substantial, audience.

Your Turn: Get Department Talks Going

OK, let's see if you can set up some department talks if they don't already exist. Follow the advice we covered and see if you can nominate someone to own the organization of talks in your department. If not, and if you've got the time, then give it a go yourself.

See if you can get a pipeline of talks lined up from your colleagues, and of course, you should nominate yourself to give one too. Maybe you could revisit lightning talks that were unanimously well received and see if they can graduate into a longer format.

One of the previously mentioned ideas for talks was the concept of horror stories—when everything goes wrong. These talks are interesting because others get to learn from mistakes without having had to make them themselves. Mistakes offer opportunities to improve how people work together rather than just committing a fix and forgetting about it. We'll have a look at how you can capitalize on these opportunities next.

Turning Problems into Learning Opportunities

Things are always going to go wrong, but there's always an opportunity to learn from mistakes and then not make the same mistakes again in the future. In this section we'll have a look at how you can begin discovering the root causes of problems in your team and how to record the outcomes. Then, we'll explore what you can do when there are bugs in processes and how they can be resolved.

Five Whys

If your team experiences an issue in their software, such as a bug in production or some downtime, then you should obviously fix it. However, an incident is an excellent chance to perform some root-cause analysis, because as a manager you should be attempting to uncover the real reasons for the problems so that they don't happen again.

A *five whys* session is an excellent way of determining the reason for an issue by repeatedly asking the question "Why?" until the root cause is uncovered. The number of whys is unimportant, but usually five is enough to get to the bottom of an issue. For example, let's assume that your application went down for a period of time and a patch was required to fix the bug that was causing it.

Before the team moves on with whatever is next in the queue to work on, get them together to interrogate the problem by asking why. Get round a whiteboard or a shared document and begin the questioning:

- *Why did the application go down?* Because the API stopped serving any requests.
- *Why?* Because the latest deploy caused a deadlock when the application started up.
- *Why?* Because the production environment has many more concurrent requests than the testing environment, and a non-thread safe data structure was used.

- *Why?* It wasn't thought about during development. The tests all passed.
- *Why?* The automated tests don't cover any high-throughput scenarios.

We hit the root cause, and the actions should be to improve the automated tests so that they incorporate a load simulation of the live environment. As a manager, performing five whys has a number of benefits:

- It determines the root cause of a problem so it can be fixed.
- It educates the whole team on what these issues are so that they can be avoided in the future.
- It reinforces that whatever happened that triggered the five whys is the kind of behavior that you want to avoid in the future. It's a signifier that something was bad enough to stop the team, think it through, and then fix the problem.

When your five-whys session is taking place, you should ensure that somebody writes up the questions and answers and the outcome. Create any tickets for actions that are required, and then share the written-up session with the whole team. Perhaps you could store all of your five-whys documentation in a shared folder so that they're easily accessible.

As a manager, you can also take the concept of five whys even further. As well as using it in reaction to critical issues in the software, you can also apply similar critical questioning to test assumptions that you or the team are making. In work as in life, it's much easier to continue doing things in the way that you've always done them rather than challenging the status quo. Because humans operate in abstractions, we can become comfortable with how things are because we don't have to think about the details. However, you should be comfortable with questioning anything that could be improved.

Why not see if there are better ways of doing things? You can apply a critical mindset to everything that you're involved in because it can produce better thinking and better results.

For example, perhaps a status quo is that running your software in the cloud will be more expensive, and as a result, you're finding that the process of deploying a new piece of infrastructure will take two months due to a hardware order rather than two days to set it up on a cloud provider. If you're facing resistance, then you can use the power of five whys.

- *Why?* We don't build applications in the cloud because it's way too expensive.

- *Why is it expensive?* When we compare the cost of cloud instances to the cost of buying hardware and paying it off over three years, it costs more money.
- *Why is that the case?* We did a comparison of running our existing infrastructure in the cloud two years ago, and the costs were almost double.
- *Which parts were most expensive?* Any instances that required reservations of SSDs or GPUs were dramatically more expensive than buying the hardware ourselves.
- *The new part of the application doesn't require SSDs or GPUs. What would that cost?* Looking at the pricing plans now, it seems that the cost of reserved instances is about equivalent to the cost of buying our own hardware. Still, we're used to buying our own hardware, so we should just do that.
- *But are there other costs with buying our own hardware?* I guess there's the time that it takes for us to build and rack the new machines—we could be doing something else instead, plus the extra power consumption in the data center, and the additional spares we'd need to keep as well. We'd also need to deal with anything that breaks, such as replacing drives or PSUs.
- *What do you think about using the cloud for this new part of the application now?* Actually, it's not that bad. We should think about how we can save time and effort by putting more stateless parts of the application in the cloud.

Try to break decisions—especially those that are contentious—down to their root and then see whether they still hold true. The simple application of these questions could save real time and money and produce better outcomes. Better still, you're demonstrating to your team that the application of *first principles* and common sense prevails in most situations, and that's exactly the behavior you want to champion.

Your Turn: Five Whys

Next time you have a production issue, perform a five-whys session. Get to the root cause of the problem, document it, and follow up on the actions.

See if you can apply the five-whys technique to other decisions that are being made in your team, such as whether to write a new piece of functionality as a service, or whether to use an existing database or to create another. See if you can make asking "why?" an integral part of how your team approaches problems.

Management Bugs

Here's a neat idea that you can begin with in your team but potentially radiate outward to the whole department. We've just looked at what happens when there are problems with your software. But what should you do when there are problems with process? Perhaps your team is finding it difficult to find documentation for the architecture of the platform, or any guidance on best practice for how to create new services. These are concerns that are wider than just the team and may be difficult to find an answer to. These sorts of concerns also begin to hamper morale: how can people effect change when it's out of their control?

You could try a *management bugs* initiative. *When eBay went through rapid expansion [Sco17]*, leadership became increasingly distant from the staff on the ground. A small, nimble culture was becoming corporate, fast. The CEO wanted to rectify this problem and therefore installed a suggestion box in a busy area of the office. Anyone could drop a note into the box, and at the all-hands meeting the box would be emptied. All questions and suggestions would be read out and answered in front of the whole company. This was a neat way of breaking down the typical barrier to leadership: all input was treated equally, no matter who it was from.

This initial implementation is good but has some problems:

- *Box location.* A physical suggestion box doesn't work well for geographically distributed companies.
- *Presence at the meeting.* If answers are read out at a meeting, then it's likely that not everyone will be able to attend, and this adds overhead to recording and distributing the talk.
- *Tracking issues.* Not all of the issues raised are going to be solved instantly. Some may require a lot of discussion, or may spin off into a working group before a solution is found. How can staff know the ongoing status of everything that's been raised?

Why not implement this management suggestion box as part of your ticket-tracking system? That way, all of your engineers can see issues that have been raised, anyone can comment, and all of the work related to those issues can be done in the open. This may seem like a left field idea, but it's something that we've had a lot of success with at our own company. Given that we have this process in place for software bugs, why not try the same for other types of issues?

Assign someone to be the main point of contact for management bugs. It could be you to begin with. The workflow then looks something like this:

- Anyone can raise a "management bug" ticket with their description of the problem.
- The ticket is then picked up and moved into whatever workflow you want to use for it. Kanban works well.
- Comments can then be solicited from anyone interested, and then actions can be decided.
- When the ticket is eventually completed, a summary can be written of the issue and the actions that were taken. This summary can be broadcast to the wider department if it's relevant.
- The ticket is then set to "done" and an archive of the summary can be put into a shared folder where everyone can see it.

Although simple, we experienced a strong uptake in the initiative at our own company. For example, here are some of the tickets that were raised, to give you an idea of the types of issues and their resolutions:

- Clarification on how perceived performance links to salary increases at the end of the year. The algorithm for how salary increases are calculated according to performance and budget is now documented and available for all to see.
- Discussion about the perceived speed that we get work done and how that has changed over time as the department has gotten bigger. The perception that more engineers and a larger company means faster work isn't always true.
- Concerns about teams switching to new projects too quickly after their completed project has shipped, thus not getting the chance to iterate, fix bugs, and address technical debt. As well as being mindful of the issue, we now try to give each new feature defined time to soak in production before context-switching on to new endeavors.
- New roles in teams were being advertised externally on the company website, but people who already work for us sometimes didn't realize that an opportunity existed for them to apply. We now make sure we advertise internally too.
- Ensuring it's clear on our careers page that part-time applicants are welcome for most positions.

As you can see, a number of department and company-wide issues were simply raised as tickets, but the power of the transparent discussion allowed the momentum to build to either fix them ourselves or to seek out others that could help us fix them. Raising tickets can be easier and psychologically safer for engineers than going directly to those that run the company. It's quick, easy, and powerful. It can even be fun. Why not give it a go?

Tools to Solve Common Problems

The final section of this chapter is to introduce you to some additional housekeeping techniques that can make good of common problems and produce documentation that you can review with time. Let's look at them in turn.

Why Is This Code Like This?

We've all been there, staring at code that's been designed in an esoteric manner. Maybe it's been your own code and you can't remember why it wasn't written differently. To prevent problems like this from happening in the future, you can use an Architecture Decision Record (ADR).[5]

As you progress on your own projects and reach a decision point—a design choice, whether that be stylistic, framework-related, or a choice on how to build some infrastructure—you can capture that choice in an ADR. The website in the footnote has links to a number of templates that you can use, but broadly speaking an ADR should capture:

- The *point in time* in which the decision was made.
- The *status* of the ADR (proposed, accepted, rejected, and so on).
- The current *context* and business priorities that led to the decision.
- The *decision* itself (what are we doing as a result of this?).
- The *consequences* of making the change, such as what is now easier or more difficult as a result.

You can store your written ADRs in a place of the team's choosing, such as in a folder on Google Drive, or, better still, perhaps you could check them in to a top-level folder in that particular codebase. Proposing a new architectural direction is then achieved by writing an ADR and raising a pull request into that codebase, where all parties that regularly commit to that codebase can see it. What's more is that they serve as an audit trail for how the code has changed over time. Even if people leave the company, the context is not lost.

You can begin a culture of creating ADRs in your team and see if they catch on within your department. They're a good subject for a lightning talk!

---

5. https://adr.github.io/

Is the Team Improving with Time?

Often teams spend so much time in the details of getting projects done that they forget to check in with how they are doing. Even if they do, it's difficult to remember whether the team is becoming more efficient, mature, and—more importantly—happy over time. That's where *health checks* come in.[6] They allow a team and those that work closely with them to reflect on a number of team health indicators and rate themselves with a traffic light system. Typically a health check happens a few times a year, and by recording the results, the team can see where things are trending with time.

Health checks are implemented by:

- Running a *workshop* session where team members discuss their current situation based on a number of perspectives.
- Creating a *graphical summary* of the result.
- Using the data proactively to *improve the situation* for the team.

The Spotify model defines ten perspectives, which each have an example of "crappy" and an example of "awesome." For each perspective there's a discussion, then team members vote as to whether they align with crappy, neutral, or awesome. Examples of perspectives to discuss are:

- Whether it's easy to release code.
- Whether the current processes are suitable.
- The health of the code base.
- The value being delivered by the team.
- The speed of getting things done.
- The clarity in the mission of the team.
- How much fun is being had.
- Whether the team is learning new things.
- Whether the team has the support they need at hand.
- Whether they are pawns or players.

For instance, an example of "crappy" for releasing code would be that it takes weeks and it isn't under the control of the team. An example of "awesome" would be that the team can deploy immediately, whenever they want.

After each perspective is discussed and voted on, the perspective is assigned a traffic light color, typically by summing up the votes in each traffic light color. Green means the team are happy enough with it right now (it doesn't

6. https://labs.spotify.com/2014/09/16/squad-health-check-model/

have to be perfect), yellow means that there are areas to proactively improve, and red means that it's bad and needs some urgent attention.

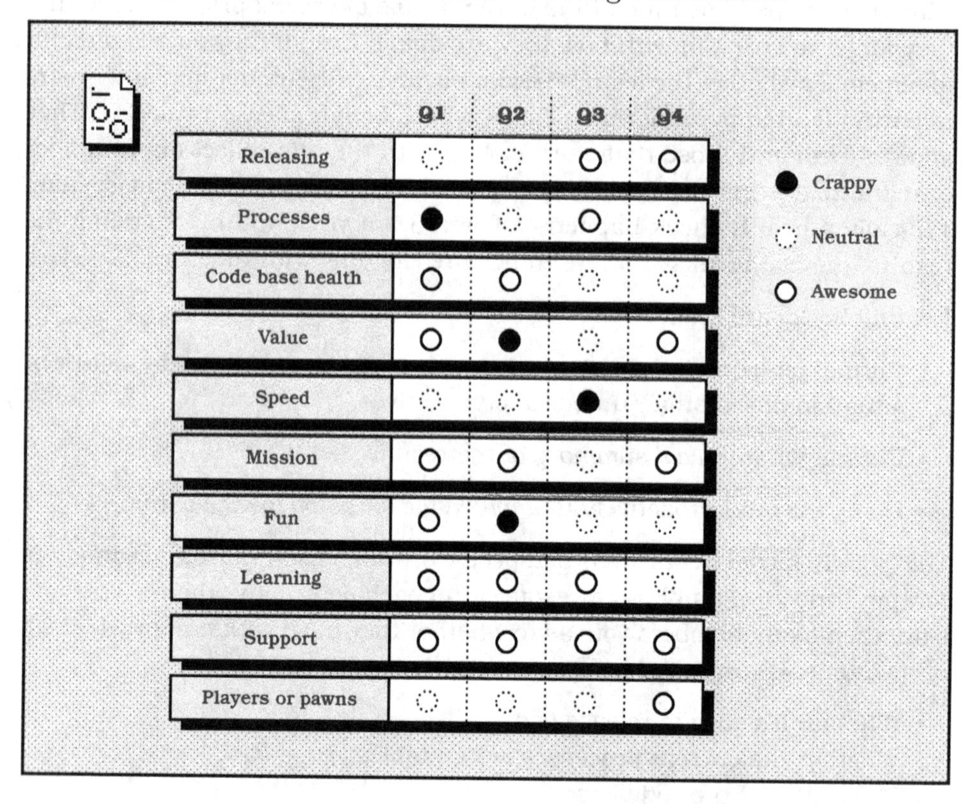

Try it out on your own team over time, maybe quarterly. Do matters trend upward? You could also give a talk about health checks and promote it within other teams at your company. If they do them, how do those teams compare with each other and why?

Who's Responsible for That?

As you know, software is complicated and so are the teams and processes that produce it. A common issue is around ownership: sometimes important things can be missed because they have no clear owner, such as a service going into production with no monitoring.

Apple addressed this accountability issue by using a concept called Directly Responsible Individuals (DRI). Typically assigned to whole projects, a DRI would be the person with whom the buck stopped. You can use this concept within your team. There will be clear DRIs for quality (for example, your QA) and for storage decisions (for example, your back-end engineer). However, you can also use the DRI principle to empower those within your team to take a stance on initiatives such as documentation, monitoring, and usage tracking—areas that are often forgotten. Think about it: could the DRIs in each team form wider guilds?

Each DRI doesn't necessarily have to be the person *doing* the work in question to achieve that initiative, but they are responsible for championing that cause and interacting with whoever they need to move it forward. It can be an empowering role.

DRIs are a stripped-back version of a traditional responsibility assignment matrix (RACI),[7] which outlines which people are either responsible, accountable, consulted, or informed for a given task or area of a project. Yet RACIs can be overkill for the needs of a team, confusing to understand, and once created, they're often forgotten about. DRIs get straight to the point when it comes to accountability. They say that a given accountability lies with this exact person, and they have whatever freedom they need to ensure it's upheld. They're easily written up and shared with others since they're just a list of names and responsibilities.

Now Let's Organize Our Careers

You've learned a whole bunch of techniques for good housekeeping. Most of these involve some lightweight process, some documentation, and some communication via broadcast or multicast.

Practicing these techniques within your team can not only improve how your team works and increase your effectiveness as a manager; it will also give you opportunities to spread your influence to others in the department by sharing the knowledge and practices in this chapter.

Here's what we covered:

- We looked at the many different forms of communication within the department and how you need to create structures in which to best harness these lines of communication. This is a pressing issue if we consider Conway's Law, which states that *organizations design systems that reflect their own communication structures.*
- We saw that you can unite people of similar skill sets and interests with *guilds*. They enable opportunities for defining best practice, bringing

7. https://en.wikipedia.org/wiki/Responsibility_assignment_matrix

people together who may have otherwise rarely communicated, and break down barriers between teams.

- We saw how you can gradually introduce a *culture of technical talks* in your team, starting small with lightning talks and working your way up to department-wide talks.

- We understood how *problems can be opportunities* for improvement, by looking at the five-whys technique for addressing technical issues, and also how starting a management bugs initiative can be empowering for anyone at any level in the department.

- We finished by looking at some *tools for solving common problems*, which were Architecture Decision Records, team health checks, and Directly Responsible Individuals.

Employing all of these techniques can begin to turn a messy department into a well-oiled machine. You should go and start implementing these! Next up, we're going to be looking at something even more important: career development. Notably, we're going to explore the two career tracks that exist in software—individual contributor and management—and go through an exercise to define these tracks so you can have these discussions with your own manager. Even better still, you can use your power as a manager to establish or improve the existing career tracks in your department, which might just be one of the most influential things that you've done yet.

Better to be at the bottom of a ladder you want to climb than in the middle of some ladder you don't, right?

> *Dave Eggers*

CHAPTER 15

Dual Ladders

It's Thursday, you've finished work, and you're sitting in a coffee shop just down the road from the office. Charlie is in town to attend a conference, and you haven't seen each other since college. Your drinks get placed on the table.

"How's work going?" you ask. "When we emailed each other a couple of months ago, you said you were working on some pretty interesting sounding streaming APIs."

"Kinda good and bad," Charlie replies.

"Why's that?"

"Well the good news is that I've finally managed to get a promotion," says Charlie, looking down at the table.

"Oh, that's great! Congratulations!" you reply.

"I'm managing a team now."

"Even better! Well done. But what's bad about it?"

"Well," says Charlie, "I'm not sure if I actually want to be a manager. But it seems to be the only way that you can get promoted without having to wait about ten years."

"Oh, so what happened?"

"My previous manager left and they didn't have the budget to hire someone externally, so they promoted me. It all happened really quickly. I don't think I knew what I was getting into."

"Have you had a chance to talk to your manager about it? What do they think?"

"Dunno. My manager is in another country now, and I just don't feel like we get to talk about anything beyond the surface level of what's going on with our projects."

You take a sip of your coffee. "That really sucks, I'm sorry. I'm still finding my way in my new job, but I'm always happy to listen."

"I appreciate that. Thank you. I just don't know whether I've made a choice I can't back out from. The team seems to like the fact I'm doing the job, but it's just making me feel like I get nothing done. Work isn't really fun anymore. I miss building things."

"Yeah, I hear you," you reply.

Way back in the *Introduction*, we stated that there was a management-skills crisis in our industry. Since we have a skills crisis, and since we expect our managers to define career progression for our industry, it could follow that we have a career progression crisis also. The world is full of companies that haven't defined what our career progression should be, causing all sorts of anti-patterns:

- Good engineers leaving their jobs to get promotions because they were unable to progress at their current company.
- Engineers reluctantly becoming managers to get promotions.
- Managers leaving companies as the solution to being unhappy in their roles.
- Unrealistic expectations on engineers to take on management duties but also produce the same amount of code as they were before.

This chapter aims to address the career progression crisis by helping you build out a progression framework that works for you and your team. Here's what you're going to learn:

- How to define the role of *individual contributor* and what some examples of progression look like on that path in our industry.
- How to do the same for the *management track*.

With that in mind, we're going to go through an exercise to create a behaviors grid for both tracks that can act as a framework for staff to discuss their progression in their careers and make reasoned decisions about whether they would like to be individual contributors or managers. We'll then use these career tracks to *debug common career progression problems* that you may have already observed during your time in industry.

Creating this career progression framework can be eye-opening for you and your team. It also offers you another opportunity to engage with the wider department and assist in creating one for *everyone* if it doesn't already exist. Awesome! That's another opportunity for you to be impactful and influential outside of your team.

Let's get going by looking at each of the two tracks in turn. And yes, before you ask, they *are* separate jobs.

Individual Contributor Track

The first track is the path of the *individual contributor*, or IC. This is the path of technical contribution. Our ICs are the ones that get the products made, the features shipped, and the architectures implemented. Without ICs, we can't make software. This is why as a manager it's your duty to ensure that they are able to do their job well and provide them with career development opportunities.

The IC track typically has a progression that looks something like the following in a technology company. This is just an example, because every company implements their progression slightly differently. Each stage in the track is called a *level*. The levels are indicated by the letter L followed by a number.

- *L1*: junior engineer (more common outside of the United States)
- *L2*: engineer
- *L3*: senior engineer
- *L4–L5*: staff engineer
- *L6+*: principal engineer

The currency of the IC is *technical expertise* and *influence*. Typically they spend a significant portion of their time writing and reviewing code and solving problems. ICs can be engineers, designers, QAs, and more. The key trait is that they are not line managing anybody. That's a different job.

Typically the bigger the company, the more numerous the levels and sublevels. If the company consists of thousands of engineers, as do Google and Amazon, then each job title can even have multiple numbered increments (such as software engineer II, software engineer III, and so on). This can assist greatly when you need to provide meaningful progression and promotion steps at scale. Additionally, you may also get senior staff or senior principal engineers in order to create even more progression levels.

Job Title Context

A job title requires the context of the organization that it exists in. A role's weight often increases as the size of an organization increases.

For example, you could reason that a principal engineer or VP engineering at Google may carry more relative seniority and impact than a principal engineer or VP engineering at a smaller company with twenty engineers.

As an IC progresses in seniority, we expect their impact to increase. But what does impact mean in practice? An individual's impact can be observed subjectively through traits such as the following:

- *The software that they write.* Are they solving problems efficiently with code in a way that is simple, elegant, and extensible?
- *Their net contribution.* Do they predominantly progress independently, setting the standard for others to follow, or do they need a lot of support from others to get their tasks done?
- *Their collaboration with others.* Are they team players that understand the needs and sometimes contradictory forces within the team and find a way forward.
- *Their mentorship skills.* Are they a natural mentor who less-experienced engineers routinely ask for help and who improve their own skills by working with them?
- *Their influence on other engineers.* Do they bring approaches to the table that others then follow? Do they advise others outside of their own team on what we should build and how?
- *Their innovative ideas.* Are they suggesting new ways that we could be using technology to enable new features, improve software speed, or reduce the cost of our software?
- *Their influence on the bottom line of the business.* Are they involved in solving problems in new and efficient ways that make other engineers faster (for example, by building reusable architecture) or the product cheaper (for example, by rearchitecting an existing system)?

You could define many other traits as indicators for the seniority of an IC in your team, and we will look at those more as we begin to define the grids. As your ICs progress in seniority, you would expect them to be working on more complex, open-ended, and self-directed problems, involved in higher-stakes projects, and having a wider reach by mentoring others and advising other teams.

As we create our career progression framework, we'll touch more on examples of how ICs can progress within different traits and behaviors. However, if you've been in industry for a while, you should already have a good feel for what the IC track looks like. After all, you've probably been on that journey already!

Management Track

You guessed it: the other track is *management*, the track that this whole book has been about. We've also repeatedly revisited how we measure the output of a manager, which is the output of their team combined with the output of the others that the manager influences. So we've got that down, and the book so far has been full of techniques for managers to increase their output and become more effective.

Generally speaking, however, the most surefire way of a manager increasing their output is for their team to get bigger. You can only have so many direct reports before it starts to become overwhelming. Teams become too large, and the amount of time that's needed to spend with each individual becomes prohibitive. A rule of thumb is that a manager should have at maximum around seven direct reports. There doesn't seem to be any hard science that proves this theory, but if you're spending an hour with each direct report in their one-to-ones, that's already a day booked out each week.

Therefore, as managers gain more responsibility, typically they gain more people. For example, a large team can be split into two with the manager promoting another manager to run one of the small teams (which now has room to grow) while still running the other team (so they don't make themselves redundant). With time, as an area grows and grows, a manager may find themselves running multiple teams, thus increasing their output. They may have explicit accountability for a functional area (for example, a collection of infrastructure teams). Alternatively, an experienced manager could be hired into an organization to manage a division, consisting of many teams.

This book has been about becoming a manager and running one team, so we won't go into the detail of managing managers, or trying to apply for these kinds of roles. However, running larger organizations is covered in *High Output Management [Gro95]* and *The Manager's Path [Fou17]*. (Do you see yourself on this path?)

So, as a manager increases in seniority, they will typically:

- *Be increasingly effective at delegation.* As they progress up the org chart, they move from delegating development tickets to delegating the output of an entire team (or multiple teams!). This requires abstracting, defining goals and outcomes, and implementing them throughout their staff.
- *Have more people reporting into them.* The higher the position in the org chart, the more staff that branch downward underneath that particular position. This can go from teams, to multiple teams, to a division, to a whole department.
- *Own larger and more important strategic areas.* Typically a subsection of the org chart will be dedicated toward some purpose. A team may work on one specific area or outcome, but a division may work on a whole application or piece of infrastructure. The CTO runs the whole engineering department.

The job title progression may look something like this. We define levels as before, but with respect to how these roles line up with the IC track in terms of experience. It's highly unlikely that a company would have an engineering manager who hasn't had a reasonable amount of experience in industry first.

- *L3*: engineering manager
- *L4*: director of engineering
- *L5–6*: vice president of engineering
- *L7+*: chief technology officer

To create more steps of progression and increasing areas of ownership, the initial three job titles may be preceded by *senior*, such as senior director of engineering. Chief technology officer is an exception because it's the top of the management chain and typically entails being a company director (although in huge companies you can have CTOs of large divisions, but that's rarer).

Compensation

Ideally, there should be no glass ceilings in either track and no interdependencies. For a technology company, an extremely experienced individual contributor and extremely experienced manager both contribute at a high level. Thus, at a similar level of contribution, they should be rewarded similarly.

In some companies, this may not always be the case, and you can detect senior leadership bias here: a "pure" technology company may lean toward paying principal engineers more than senior managers. A more corporate environment may value leadership over technical skills, and the higher ends of the average pay bracket may favor those in management positions.

If you have easy access to the senior leadership in your company, then this is an interesting discussion to have with them. What do they value? Which skills do they think are irreplaceable?

Creating a Progression Framework

If a career progression framework isn't already available at your company, or if you would like to supplement the existing one with team-specific details, then the following exercise is a lot of fun to do. It also offers opportunities for you to increase your influence in your department, as other teams—or maybe even the whole department—can help develop it further.

Before we discover how to do it, it's worth mentioning what career progression frameworks *are* and *aren't* for. To begin with, they *are* there to:

- Help your staff talk to you about where they feel they currently are in their career and what skills they would like to work on.
- Stimulate conversation about where they would like to progress to in the future.
- Set expectations of where they might go if they continue to perform well at the company.

Frameworks should act as a compass (a general direction that allows the user to find their own way) rather than a GPS (a specific route to follow whether you like it or not). Here's what the career progression framework *isn't* meant to do:

- Rank staff against each other.
- Prevent people from progressing.
- Force anyone to progress in ways that they don't want to do.

In this section, we'll go through an example of building out a subset of a career progression framework for ICs and managers. You can then work this into a more complete framework that suits your team and your company. This will enable your staff to have easier conversations about career progression and will even allow you to have more fruitful conversations about your *own* career progression with your manager.

The Grid

A career progression framework is a grid that consists of *roles* across the top and *competencies* down the side. They should be defined in a way that doesn't make them specific checkboxes to tick to unlock a promotion. Instead, they're meant to serve as a guide to your staff and yourself when having regular discussions about progression within the company.

The Individual Contributor Track

Let's take a subset of the roles that we listed for the IC track previously: junior engineer, engineer, and senior engineer. Those go along the top of the grid. Down the side will be the competencies. But what should those be? That's up to you, your team, and your department to define them.

Here are some examples of competencies that you could use:

- *Professional experience.* How long has somebody been working in this role in general? Have they recently graduated from college or are they a seasoned professional?
- *Technical knowledge.* Are they an expert in a number of technologies, or are they beginning their journey to mastery?
- *Mentorship.* Are they receiving mentorship from others, or are they a go-to mentor for the team?
- *Conflict resolution.* How experienced are they at unblocking decisions and critiquing approaches?
- *Communication.* How is their technical and nontechnical communication? Can they clearly communicate how to solve engineering problems and also explain that to nontechnical stakeholders?
- *Influence.* Are their relationships mainly focused within the team, or do they influence decisions in the wider department?
- *Efficiency.* Do they solve problems in an efficient manner and, better still, produce reusable code and systems that speed up others also?

- Your *company values*. If your company has a set of *values*, then how would each level of seniority act in relation to those values?

The exercise to produce the grid then involves filling in the cells at the intersection of role and competency. It can be a fun exercise to do as a team. Let's look at an example of how you could fill out some of the competencies against roles below.

Competency area	Junior Engineer (L1)	Engineer (L2)	Senior Engineer (L3)
Industry experience	Has some experience with relevant technology from a course (e.g., undergraduate degree or self-directed learning). Has spent little or no time in industry.	Has 3–7 years of experience, including education. Is experienced at working in a software development team and understands how teams work together to create and ship software.	Has 7+ years of experience. Is a technical expert in their particular area and has a solid record of getting things done to specification and on time.
Technical knowledge	Is able to analyze problems and propose solutions. Should be able to implement relatively simple solutions independently and more complex solutions by asking relevant questions to more senior engineers. Balances time contributing to projects with time dedicated to learning and developing new skills.	Proficient at their chosen language and stack, including open source software and frameworks. Is able to pick up new technology and teach others how to use it. Able to tackle most problems independently.	An expert in various technologies. Possibly consulted on technical decisions from teams other than their own. Is a code merger for their team. Is knowledgeable of the latest and greatest open source software and techniques and can innovate in their team(s). Shares their knowledge with others.
Mentorship	Is likely to be receiving mentorship from more senior staff.	Able to mentor junior engineers up to their level to make them great team members.	Is a go-to person to mentor others on various parts of the technology stack. Is able to turn engineers into experts. Is a great teacher.
Influence	Is able to form strong working relationships with their colleagues.	Has a reputation of getting things done and being able to mentor others technically.	The person that everyone wants to help on their project. Others actively seek this individual's knowledge and advice to validate their approaches.

Remember that no list, framework, or track is perfect. They're an attempt to help define levels, but no one will fit perfectly. That's OK. Instead, here's what the career progression grid allows your staff to do:

- Identify areas where they are strong and areas where more development is required or opportunity exists.

- Pick hotspots or logical bite-size improvements that make sense to work on improving, then set goals with you to improve in those areas.
- Discuss how they are getting on regularly with you during your one-to-ones.

Your Turn: Create a Career Progression Grid for ICs

You could run this as a workshop with your team. What are the competencies that are important in ICs and how can those be described across the different seniority levels? Do your team's ideas match up with the existing progression framework at your company, if there is one already?

Show the results of your workshop to your manager. What do they think about it?

Discuss what progression looks like on the management track with your manager. Is it clear? Does it need defining further?

The Management Track

Creating a career progression track for ICs, especially if one doesn't exist already, is a high-impact activity that you can do with your team. Not only does it give them a framework that they can use to work out where they are, it allows them to plan where they want to go.

But what about the career track for managers? It's likely that's less important for your team right now but still important for you and others in your role at the company. You can begin thinking about it with your own manager and your peers. The existence of a defined management track allows your ICs to better understand what that role is about, and whether they would be interested in doing that role in the future. If you ask around, you may find that other managers are wondering about their own career progression too. Collectively you could have a go at defining it, creating a bottom-up movement and making some new connections along the way.

If you look back at some of the competencies that we outlined for the IC track, such as industry experience, technical knowledge, influence, and mentorship, then you could reason that these are relevant to the management track as well. The only thing that's different is that instead of employing all of these skills to build software, the managers are using them to build *people*.

You can start by copying the track that you created for ICs and deleting the content of each grid cell. Then, you start again but reason about what it means to build a team rather than software.

For example, if we take the competencies of industry experience, influence, and efficiency, we may form a management track that looks like the following for the engineering manager, director of engineering, and VP engineering roles.

Competency area	Engineering Manager (L3)	Director of Engineering (L4)	VP Engineering (L5)
Industry experience	The individual has made the conscious choice to become a manager. Could either have had previous management experience, or wants to get started on the management track. The company can offer training and mentorship to first-time managers. Has excellent technical skills and interpersonal skills and leads from the front. Is able to communicate well with Product, Engineering, and Design to ensure that projects ship successfully.	Has a proven track record as a manager of people and teams and of completing projects. Accountable for a subdivision of the engineering department, potentially with multiple teams (e.g., a strategic area, technical function, or geographical area).	Has proven experience managing multiple teams and owning a strategic or technical area. Runs a division of the engineering department, consisting of many teams.
Influence	Leads a team from the front, delegating to empower others. Is someone that others want to work for. Makes their team a better place. Influential for their communication skills as well as their technical ability.	Influences all staff in their division. Ensures that their teams are efficient and deliver great projects. Debates and discusses future technical and product direction. Works with the VP engineering on the strategy of the division.	Influences the department both inwardly and outwardly. Ensures that their teams are efficient and deliver great projects. Debates and discusses future technical and product direction. Key stakeholder in how and why we do things in the whole department.
Efficiency	Delegates work effectively amongst their team. Knows that the output of their team will always be greater than their own output. Will debate prioritization of tasks to get things done more efficiently. Is able to organize their time to be effective in their role.	Utilizes their teams to delegate work effectively. Is able to debate and decide what and how we should be doing to move us toward being a better division. Chooses to spend their time in the most impactful areas.	Builds upon the competencies of a director of engineering to oversee the output of the whole division. Works closely with their staff to debate, prioritize, and decide the best path forward.

By keeping the competencies that are required as similar as possible, it allows staff to better understand the differences in focus and output that the company expects between ICs and managers.

However, it's also perfectly normal to have unique competencies on either track. For example, it may be useful to have the typical number of people below a given managerial position in the org chart (that is, an engineering manager will have one team, but a VP engineering will have an entire division). Likewise, it may be beneficial to outline some ballpark estimates of the percentage of time ICs spend between writing and reviewing code and reviewing designs and architectural approaches. It's up to you: whatever helps your staff identify with the characteristics of given roles is useful to include. See what others think should be in there.

Your Turn: Create a Career Progression Grid for Managers

Talk to your manager to find out if there's a career progression framework for managers in the department. If there isn't, work with them and your peers to create one.

You can run this as a workshop session that will also give you the advantage of forming connections with your peers.

If you're in a small company and don't have many (or any!) peers, then why not do this exercise to define what it should be in the future if your company grows?

Career Progression Troubleshooting

Even if you define detailed career tracks for both ICs and managers, people will still find lots of bugs. You might even be experiencing some of them yourself as a new manager. Let's do some debugging and see if we can find some answers to common questions.

Isn't a Manager on Both Tracks?

If you're a manager, then you might be thinking "Hang on, aren't I on both tracks? I code as well!" It's a good question. There are plenty of managers that are still able to produce high-impact code contributions as well as manage a team of engineers.

However, the track that you're on defines your *primary function.* If you're on the management track and still able to produce a lot of code *without putting your managerial responsibilities at risk,* then that's fantastic. However, if you're on the management track and you are coding to the detriment of your managerial responsibilities, then you need to rethink how you're spending your time.

As you'll have seen by reading this book, becoming an effective manager takes a lot of practice. If you excel and become a natural at management, then you may find yourself able to free up more time for your own code contributions. However, you should make sure that you're performing well as a manager first. Your team need you to do your job.

What If I Don't Like Management After I Try It?

Since we've defined that the IC track is different from the management track, and hence that they're different jobs, we can better understand why engineers are hesitant to make the leap into management. What if they do so and then don't like it after all? Do they become stuck and then have to leave the company if they don't like it?

It's possible to create a safety net. Imagine that one of your staff has applied to run another team at the company—and you know that they would do a stellar job—but they're worried about what will happen if they don't like management.

Before they start, encourage them to create a *30-60-90-day plan* between themselves and their new manager. The process is simple: they should define a number of objectives to achieve within each of those time periods. It can be as granular as is useful. Examples of these goals could be completing a formal handover of staff, beginning weekly one-to-ones, deploying a fix to production, and so on.

You can bake a get-out clause into the plan. Since managers only experience the role for the first time when they actually do it, the clause can state that if at any time it isn't working out, they can return to their old role with no hard feelings whatsoever. This way you prevent forcing a high-performing employee into leaving because they've tried to progress their career but have ended up in the wrong role.

The Peter Principle

The Peter Principle is a management concept coined by Laurence J. Peter. It observes that in a hierarchical organization, employees are promoted based on their previous success until they reach a level that they are no longer competent.

Instead of falling foul of the Peter Principle, making both your employees and organization unproductive, this chapter has detailed a career progression framework so that staff know how to develop in a way that suits them.

Additionally, by handling transitions into managerial roles with a get-out clause, you can avoid managing high performers out of the organization unintentionally.

Isn't Progression Subjective?

Yes, absolutely. Career progression frameworks should not provide discrete checkboxes that employees have to fulfil to move to the next level. Not only is it nigh on impossible to empirically measure productivity and impact, it potentially opens all sorts of back doors to game the system and act unethically.

As you spend time as a manager, it will become apparent when people are ready to progress. You'll notice their competency and expertise increasing. Their peers will rate them highly. Typically, a promotion should be a ratification of a position that has been almost reached. No promotion should be a surprise. An engineer progressing to Senior should be a promotion that is self-evident to their peers. If it isn't, then you've probably made the wrong call.

Because of the subjectivity of progression, you should be regularly getting peer feedback as part of performance reviews as we outlined in *The Most Wonderful Time of the Year*. As time progresses, an increasing collection of peer reviews combined with both your and their judgment of their performance against a career progression framework should make it clear when people are at the right stage to be promoted.

Promotion is not a leap of faith. Think of a promotion as the first night on stage after years of practice, rather than being thrown out in the spotlight with no cue and no clue.

What If There Are Big Variations Between Skill Sets?

If the creation of the career progression framework begins to catch on in the wider department, it's likely that roles with subtly different skill sets will find that it doesn't quite work for them. For example, a sysadmin may find that the IC track definition doesn't quite cover their own progression.

Two actions can be taken to resolve this issue:

- Further genericize the main career progression framework so that it's applicable to as many skill sets as possible.
- A skill-set group can produce their own overrides or extensions of the career progression framework. For example, they may make certain definitions of competencies more specific, or they may add new competencies to it.

Depending on the size of the company, career progression frameworks may be broken down by functional areas of a department, such as infrastructure or mobile applications.

How Do Career Tracks Decide Compensation?

The short answer is that they don't explicitly decide compensation, other than identifying promotion boundaries that should be tied to a compensation increase. Progression along a track is a career's work. It's important that the organization acknowledges that being promoted into a new position is not the only way that employees can receive meaningful raises. For example, high-performing individuals should be able to increase their compensation at a faster rate than those in the same role performing as expected.

Instead, career progression frameworks are just another variable that should be used when it comes to deciding compensation. Your company should already have some kind of framework for compensation already in place, but if they don't, then the following algorithm can act as a rule of thumb for deciding how to increase individuals' compensation in what is often a zero-sum game (that is, there's fixed money available to increase a group of individuals):

1. Is anyone underpaid according to market rate or their peers within the company? Fix this.

2. Is anyone being promoted into a new role? Apply the increase from the promotion to market rate.

3. Apply whatever the decided inflation and cost of living increases are for a given location to everyone not affected by the first two steps.

4. Then, increase pay in line with performance within the bounds of the remaining budget. Given that a fixed percentage will increase a larger salary more than a smaller salary, ensure that less-experienced staff who are more rapidly increasing their value to the company get higher-percentage increases to counterbalance this effect.

If you are unsure how pay increases are handled at your company, then you should ask. It's an excellent conversation for you to have with your manager.

This Doesn't Work for My Company!

Does it not? That's interesting. Why is that? Is it that your company has a progression structure that's overly complicated? What do people think about it? Are you facing resistance, and if so, why? Has management not been identified as a different role to IC? Are the paths to progression unclear or nonexistent? Would the company rather not think about this issue? You can change this with support from your peers.

A number of examples of career progression frameworks are available to view online.[1,2] These can give you more ammunition to prove that career progression frameworks are well worth spending the time defining, both for your staff and for yourself.

The two tracks we've outlined are simply a model: they're a great place to start if you've never thought about or formally defined career progression before. Other models also exist, such as a three-pronged trident of management, technical leadership, and individual contributor.[3] Staff may even benefit most by swinging often between tracks, improving their collective skills.[4] Your mileage may vary.

Your Turn: Score Yourself!

With the career framework that you've created, or by using the framework of another company that you've found online, score yourself against each individual competency by using the level you think that you are.

For example, you may rate yourself at L3 for communication, L4 for industry experience, and L5 for technical knowledge.

You can also do this exercise with your staff as part of their review process. We'll look at how to do that later in the book in *The Crystal Ball*.

Time to Tackle the Big Issues

So there you have it: you now have the tools and the motivation to define career progression for you, your team, and maybe even the whole department. If nothing like this exists at your company, then I encourage you to start this movement at your workplace. You should start by defining a progression framework with your team as an exercise to define how *they* see their roles and your own role. Then you can begin to invite others to see what you've been working on and see whether the network effect can bring about real change.

Career progression frameworks *lower the barrier* to enabling career progression discussions. They bring specificity to the skills and competencies that ICs and managers should be aiming toward, and enable those who've been too

1. https://www.progression.fyi/
2. https://www.levels.fyi/
3. https://www.thekua.com/atwork/2019/02/the-trident-model-of-career-development/
4. https://charity.wtf/2017/05/11/the-engineer-manager-pendulum/

reserved to ask about what their future holds to have those conversations with greater ease.

Here's what we've covered:

- We defined the role of *individual contributor* and considered what progression looks like on that track.
- Then we considered progression on the *management track*.
- Using the example progressions from those tracks, we saw how you can define *competencies* that can be mapped against roles. This allows you to create a career progression framework that is adaptable to your team and department.
- We then looked at some *common bugs* that arise around career progression frameworks and what you can do about them.

In the next chapter, we're going to dive deep into an urgent and critically important issue with our industry: inclusivity and diversity. This is a crucial area in which we, as managers in technology, need to unite and make an impact.

Someone's sitting in the shade today because someone planted a tree a long time ago.

> *Warren Buffett*

CHAPTER 16

The Modern Workplace

Sarah—the CEO—starts walking toward your desk looking flustered and stressed.

"How's the board meeting going?" you ask.

"Fine. I'm going to be showing the board around our office in a moment. I need you all to be busy working hard. Some of you should get some code up on your screens. A couple of you should get in front of a whiteboard and work on a problem."

"Huh? What sort of problem?" asks Tara.

"Just anything. Draw a diagram of some of the platform. Point to it. Ask questions about it."

"What?"

"Just do it! We need to give them a good impression." Sarah spots something across the room. "Wait, what are you doing?"

Paulo is packing up his bag. "I do the school run on Wednesday so I leave a little earlier."

"Not today you don't," says Sarah. "We need to be at our desks, working hard. Hustling. The board need to see that we really want this."

"But who's going to pick up my daughter?" replies Paulo.

"It doesn't matter. I need this from everyone today. Nobody leaves until the office tour is done. We need to show that this company matters more than anything right now!"

You and your team fall silent. Sarah jogs back to the board meeting.

"What the hell?" you cry. Paulo is making a frantic phone call. Tara is drawing an architecture diagram on the whiteboard. You look around to see people regrouping at their desks.

"What am I even going to talk about?" says Tara, pointing at the diagram. She assumes the posture of a weather presenter with arms outstretched. "Hmm, yes! This is our database! It is a good database. I love it!"

"Have you ever seen them in person before?" asks Ben.

"I haven't even seen their pictures," replies Tara.

One by one you watch the board emerge. First up, a white guy in his fifties. Blue jeans, white shirt. Next, another white guy in his fifties. Blue jeans, white shirt. Next, a white guy in perhaps his late fifties. Black jeans, blue shirt, grey fleece. Then come two more fifty-ish-year-old white males in shirts. One is wearing a pair of Converse. One walks over.

"Hey, I'm Wayne. Wayne Hunter. Former CEO of Globex Corporation. I'm one of your board members."

"Hey, Wayne," you all chime in unison.

"You know, I love seeing you guys. Going for broke, making it happen. I remember the early days at Globex when we just started out. Eighteen-hour days, working through the night, grinding hard. Sleep was for losers!"

"Yeah, I guess it's exciting work," Tara replies.

"You betcha. Nothing better than the grind. Making a success of yourself. Missed the birth of my son because I was raising a VC round. We went hard!"

"Erm, OK," Ben says. "But how did your wife feel?"

"I have a strong woman who takes care of business at home so I can take care of work," remarks Wayne. "That's what strong American couples do!"

Blank faces. An awkward silence.

Tara turns and raises her hand to the box diagram with her palm outstretched. "So, would you like to hear about our database?"

As a manager in a fast-moving industry, you're going to be expected to understand the pressing issues that we face. Being able to do so will make you a better, more understanding manager who will be able to support and nurture a strong and diverse team and also fight back against prejudice, stereotypes, diversity, and pay imbalances, and you'll contribute to making your company a better, more forward-thinking place.

After all, we all want to work for companies and managers that care, understand, look out for us, and fight our corner. By exploring the topics in this chapter and doing further reading on your own, you'll be able to do the same. We're going to cover a multitude of areas, but they are fundamentally linked by one core principle: that each and every one of us, regardless of our gender, color, or background, deserve to be treated fairly and equally and have the same opportunities to succeed as everyone else.

Here's what we're going to cover:

- *Diversity and inclusion* in technology, which covers a multitude of facets such as gender, color, age, background, and the way in which we work together and thrive. We will look at some industry statistics and pressing concerns and inform you as to how you can change the way you think about these issues and what you can do to encourage positive improvements.
- *Flexible and remote working*, which not only allows you to give your staff the ability to better balance their work and home lives, but also allows you to recruit globally and thus increase your diversity and talent pool.
- *Work-life balance*, notably examining the popularization of *hustle culture* and the ill effect that it's having on our industry.
- *Culture*. What is culture? How do you get a good culture, and what can you do about it?

After you finish this chapter, not only should you be well-versed to engage and learn more about these issues, you should also be able to rate how your current workplace fares and consider how you can actively make it a better place.

Diversity and Inclusion

Diversity is about the individuals that make up your workforce. *Inclusion* is about how you encourage and enable a work environment that allows all employees to be psychologically safe and able to contribute. To begin with, let's look at diversity.

If you didn't already know, our technology industry has a diversity problem. Published diversity statistics[1] show that despite the 2017 U.S. population being 51% female, at the time Nvidia only employed 17% female staff, Intel and Microsoft 26%, Dell 28%, and Google, Salesforce, and YouTube 31%. This reporting also didn't account for those that identify as nonbinary or transgender or the fact that the diversity gap widens at the most senior levels of companies: a 2018 report found that only 10% of tech executives are female.[2]

But the diversity problem goes beyond just gender. Racial diversity in technology is poor, and even less is being done about it. Consider *colorless diversity*: a phenomenon where the industry is not investing equally in addressing the imbalance of people of color. Due to innate human bias and lack of awareness, predominantly white males with the power to make hiring and promotion decisions subsequently tackle the gender imbalance issue by focusing on white females.[3]

We have every reason to want a diverse workplace. Research shows that diverse groups perform better than homogeneous teams. A 2015 McKinsey report on 366 public companies found that "those in the top quartile for ethnic and racial diversity in management were 35% more likely to have financial returns above their industry mean, and those in the top quartile for gender diversity were 15% more likely to have returns above the industry mean."[4] In addition to financial metrics, diversity within R&D teams *generates certain dynamics that foster novel solutions leading to radical innovation [DGS13]*. Regardless of the statistics that try to quantify *why* we should have diverse teams, we should have them simply because *humanity is diverse*, and settling for anything less is prejudiced. That's 100% true.

That's the upside. But there are big, bad downsides to not focusing on improving diversity. For example, a lack of awareness of racial bias can mean that datasets used to train image recognition software don't have sufficient representation from non-white groups, thus producing webcam software that can't recognize black faces[5] and, even worse, that automatically categorizes black faces as gorillas.[6]

---

1. https://informationisbeautiful.net/visualizations/diversity-in-tech/
2. https://www.entelo.com/press-releases/quantifying-the-gender-gap-in-technology-entelo-reveals-analysis-of-450-million-candidates/
3. https://qz.com/523137/the-tech-industrys-diversity-focus-favors-one-group-over-pretty-much-any-other/
4. https://hbr.org/2016/11/why-diverse-teams-are-smarter
5. http://news.bbc.co.uk/1/hi/technology/8429634.stm
6. https://www.theverge.com/2015/7/1/8880363/google-apologizes-photos-app-tags-two-black-people-gorillas

Worse still, a lack of diversity sustains bad workplace cultures, power asymmetries, and exclusionary hiring practices, buries harassment, and enables unfair compensation practices and tokenization that causes staff to leave or avoid working in the industry altogether, thus highlighting how diversity and inclusion are closely linked, even though they are separate.

Discriminating Systems: Gender, Race and Power in AI [MWC19] is an eye-opening piece of research on the subject of diversity in the AI industry which is applicable to the technology industry as a whole. The findings are revealing and shocking. The authors, at the time of writing, found that women comprise only 15% of AI research staff at Facebook and 10% at Google. For black AI workers, there are only 2.5% at Google and 4% at Facebook and Microsoft.

But what good is improving diversity if the culture of the company isn't inclusive? Without an inclusive culture, improving diversity is just a numbers game. Improving diversity is akin to more people getting an invite to the dinner party, but it's inclusion that gets them a seat at the table. As a manager, you need to understand these issues and start taking concrete steps to improve it. Let's look at diversity from two different aspects:

- The *pipeline* of candidates that come from education into industry (often *overrepresented* in its contribution to the problem, mostly affecting *diversity*).
- *Cultural issues* at organizations (often *underrepresented* in its contribution to the problem, mostly affecting *inclusion*).

Pipeline

Clearly one way to increase the diversity of an organization is to hire more diverse people. The word *pipeline* is often used to refer to people moving through the education system to the point that they're able to be hired into industry, typically at the end of a university degree program. This is the first point where we begin to see problems.

In 2015, only 18% of computer science bachelor's degrees awarded in the United States were to women.[7] This is in stark contrast to the overall science and engineering bachelor's degrees awarded, where *half* went to women. The proportion of degrees awarded to black women *declined by 11% between 2000 and 2015 [Nat18].*

7. https://ngcproject.org/statistics

It's not entirely clear why this happened. However, data shows that in 1984 the percentage of women studying computer science began to decline, and has been declining ever since.[8] The same pattern was not observed in other technical and scientific fields. One theory is that the decline began at the same time that personal computers began to enter the home, which were marketed almost entirely to males.

In the 1970s, computer science degrees assumed no prior knowledge. However, after the personal computer revolution, that changed. Professors began to assume that all students had computers at home, creating a gender and class divide. On college campuses, computer science began to be housed in separate departments,[9] further separating it from other disciplines with a more even gender split such as mathematics.

With fewer non-white, non-male candidates becoming available to the job market, it becomes harder to redress the balance. In a recent study, young people report being connected to the internet at rates of 94–98%.[10] But despite this, those of a lower socioeconomic status are less likely to have opportunities to apply their digital literacy.

Yet the pipeline of diverse candidates isn't the only problem.

Cultural Issues

Women constitute 24.4% of the computer science workforce and get paid a median 66% when compared to men. Legislation in the United Kingdom for companies to publicly report these gender pay gap figures has reinforced this statistic for many companies: the U.K. branches of Civica, Huawei, and Siemens paid women around 40% less than men in 2018–2019.[11]

This doesn't necessarily mean that for the same roles and experience women are being paid less than men (although it does happen and is shameful). Typically what this means is that while males still dramatically outnumber non-white males at the most *senior and well-paid positions*. Note that if a company recruited tens of female graduates the following year as a response to the gender pay gap issue, then the gap would likely *widen*, as those new recruits would be at the lower end of the pay band. This can be represented diagrammatically.[12]

---

8. https://www.npr.org/sections/money/2014/10/21/357629765/when-women-stopped-coding?t=1577101365400
9. https://www.goodcall.com/news/women-in-computer-science-09821/
10. http://www.civicyouth.org/PopUps/WorkingPapers/WP59Kahne.pdf
11. https://www.theregister.co.uk/2019/04/04/pay_gap_figures_201819/
12. https://yourlossbook.com/the-attrition-triangle-model-diagram/

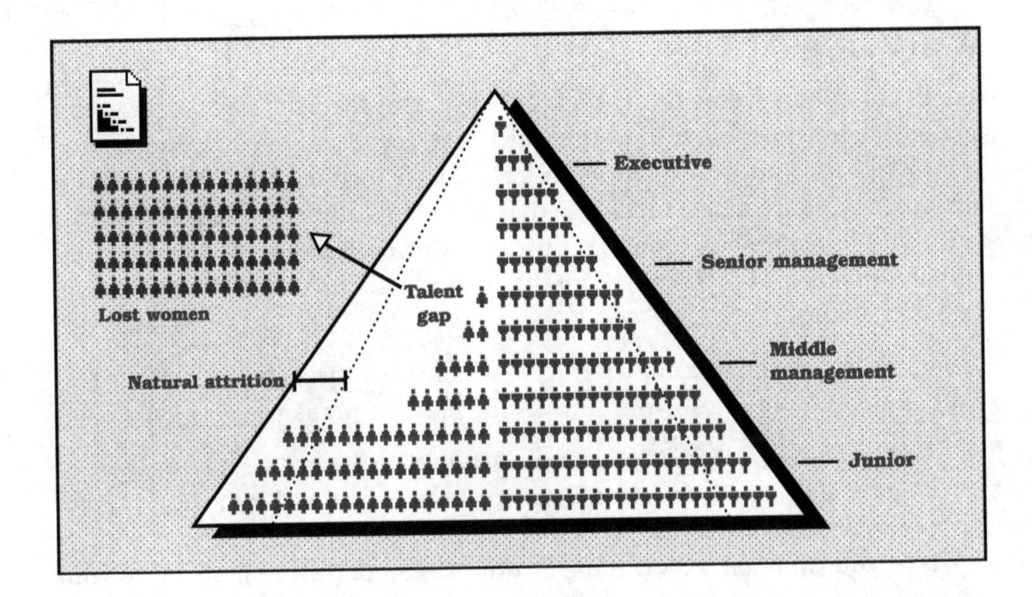

Our industry has a lot of work to do to attract and retain a diverse range of candidates. However, the reality is, unfortunately, worse than that. For example, a survey of thirty-two leading technology companies found that out of $500 million in total philanthropic giving, only $26 million went to programs for outreach to women and girls, and just $335,000 specifically toward women and girls of color.[13]

Many technology companies have also done a poor job of creating the kinds of inclusive environments that foster diversity. A case has been argued that outward focus on *diversity* initiatives has been a Band-Aid over larger internal *inclusion* issues such as sexual harassment, discrimination, racism, and misogyny. In 2018, thousands of Google employees walked out in protest against the impunity of senior leaders accused of sexual misconduct,[14] prompting similar action at other companies, such as Amazon.

Disability is also a key part of diversity and inclusion. According to 2018 U.K. data, even though disabled people had increased their employment rate from 43.4% to 53.2%, there was a reported disability pay gap of 12.2% across all industries.[15] Many employers are not well educated on interviewing people with disabilities or providing workplaces that are suited for access for those who are not able-bodied.

13. https://www.rebootrepresentation.org/report-highlights/
14. https://www.ft.com/content/af5f42a8-1501-11e9-a168-d45595ad076d
15. https://www.ft.com/content/5cca6580-150b-11ea-9ee4-11f260415385

What You Can Do

So, the situation seems pretty bad, and it is. But the good news is that as a manager, you now have more political and social capital to educate and inform others and also make your company a better place through your actions. If you are a new manager, then it's likely that you will not have the authority to make these changes outright. However, as you learned in *How to Win Friends and Influence People*, you can lean on your network to begin to enact change.

Before we look at the long-term strategies, remember these three key principles that should be present in your day-to-day life as a manager:

1. *Remember your bias.* We all have an implicit bias,[16] deeply engrained as part of our culture. We all act on the basis of prejudice and stereotypes without intending to do so. If you are a white male, it's likely that your implicit bias will have you hiring more white males into your team, or unconsciously judging a forthcoming male as a *leader* while judging a female exhibiting the same characteristics as *pushy*. Be mindful of everyone's differences because everyone has something to contribute.

2. *Call out incorrect behavior.* If you happen to see or hear casual or discriminatory sexism, racism, ageism, homophobia, transphobia, or anything else untoward, then you *call it out*. You do not need to confront people in the moment. Instead, you can request a few minutes of their time privately and explain why what happened was unacceptable to you. If it was a serious incident, you get your manager or their manager involved.

3. *Seek to champion inclusiveness.* Ensure that the way in which your team works together is supportive of those that need to work flexibly and across multiple locations, seeks to consult and listen to all employees, promotes courtesy and respectful communication, and is mindful of individual needs.

Leading a no-tolerance approach by example will soon make that no-tolerance approach spread to those you work with and those you influence.

16. https://plato.stanford.edu/entries/implicit-bias/

Positive Discrimination Versus Positive Action

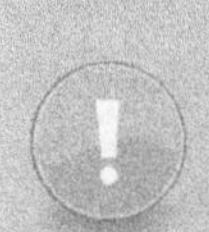

Positive discrimination is illegal. Positive action is legal. They are very different. Consider a situation where a candidate is being interviewed.

If an employer gives preferential treatment to candidates from under-represented or disadvantaged groups regardless of their ability to do the job, then that's positive discrimination. They must always consider the abilities, merits, and qualifications of all candidates in the exercise.

However, an employer can take positive action if they think that an equally qualified candidate from an underrepresented group suffers a disadvantage compared to those that do not, and can take it into account when choosing who to recruit or promote.

Support the latter. Prevent the former.

Now, let's look at strategies to increase workplace diversity and inclusion that will require you to use your network and lobby for greater change in your company. Given the force of inertia and the sheer size of some of these actions, you might not succeed in achieving them. However, the work that you and others put into pushing toward them will move the needle, and it's important that it is always moving.

These suggestions are a slightly adapted version of the recommendations from *Discriminating Systems: Gender, Race and Power in AI [MWC19]*:

1. *Publish compensation levels*, including bonuses and equity, across all roles and job categories, broken down by race and gender.

2. *End pay and opportunity inequality*, and set pay and benefit equity goals that include contract workers, temps, and vendors.

3. *Publish harassment and discrimination transparency reports*, including the number of claims over time, the types of claims submitted, and actions taken.

4. *Change hiring practices to maximize diversity*, including targeted recruitment beyond elite universities, ensuring more equitable focus on under-represented groups, and creating more pathways for contractors, temps, and vendors to become full-time employees.

5. *Commit to transparency around hiring practices*, especially regarding how candidates are leveled, compensated, and promoted.

6. *Increase the number of people of color, women, and other under-represented groups at senior leadership levels* of companies across all departments.
7. *Ensure executive incentive structures are tied to increases in hiring and retention of under-represented groups.*

Your Turn: Get Real About Diversity and Inclusion

We have a crisis, so now is the time to act. Does your company already have a diversity and inclusion committee or similar? Get in touch with them. If they don't, can you assist in starting one? Educate others about the issues that our industry faces and what we can begin to do to fix them.

Every single day, live by the two principles: remember your bias and call out incorrect behavior. When candidates are interviewed, is it by an all-white-male panel? Fix that. Are diverse participants overlooked for department-wide talks or important initiatives? Fix it. Is your company biased to listen to physically present employees more than remote workers? Change your practices.

As you progress in your management career, remember that the opportunities that you give to others are more important than the ones you get for yourself: you may already be privileged. Find, train, and promote people from diverse backgrounds. There is no excuse. You shape the future of our industry.

The Shift Toward Remote Working

The internet has revolutionized the delivery of software, and it's also revolutionizing *where* we create it. The proliferation of high-speed internet connections and collaborative software for doing our jobs is creating opportunities to shed the burdens that faced previous generations: the need to live a commutable distance from a busy city, the need to spend hours daily on commuting itself, and the preconception that the only way in which high-quality collaborative work is done is when humans physically exist in the same space.

At the time of writing, numerous notable software companies are fully or mostly distributed worldwide. At Mozilla 46% of their 1,000 staff members don't work in their offices;[17] 85–90% of Hashicorp's 800-plus staff work remotely;[18] and 100% of Automattic's 1,150-plus employees are remote.[19]

It isn't just fully remote companies that are beginning to change the way in which they work. The demands of everyday life, especially when children or

17. https://blog.mozilla.org/careers/working-on-distributed-teams/
18. https://twitter.com/mitchellh/status/1201543638842408960
19. https://remote.co/company/automattic/

elderly relatives are present, combined with the current ease of being productive away from the office, are forcing an evolution in how even traditional companies operate. It's common to have employees working more flexible hours and working designated days from home.

You may already be used to working with a distributed team. However, supporting remote workers *well*, especially as their manager, is a learned skill. We'll be looking at some of the nuances in this section. Remote working is a subject that has numerous excellent books written about it, so we'll only be skimming the surface. If you're interested in seeing more about how a remote-only company documents their culture and processes, then you should check out *Gitlab's Handbook [Git19]*. If you're looking for a deep dive into the culture shift of remote working, then check out *Remote: Office Not Required [FH13]* and *The Year Without Pants [Ber13]*.

It's Not as Easy as It Looks

Being able to work from anywhere benefits both employees and employers. Employees gain more flexibility over where they live and how far and often they need to commute. Typically, fully distributed companies employ staff over multiple time zones which means that typical office hours begin to shift and communication becomes more asynchronous. Employers dramatically widen their talent pool by being able to hire from all over the world, reducing competition for staff in small geographical areas and increasing workplace diversity.

Great, huh? Mostly. It isn't straightforward. Adapting to support remote working is a culture and mindset shift that many companies find too challenging. It requires more effort in communication and documentation, a rethink around interruptibility and availability to others, and the use of clear priorities and goals to focus staff so they can operate independently rather than relying on interactions around the office.

It's unlikely that existing non-remote companies will pivot to being completely remote. Many people are happy with the status quo. However, it *is* likely that we'll see more companies becoming *hybrids*, where part of the workforce works out of company offices, part of it is fully remote, and the remainder split their time between both.[20] This is the future that you need to prepare yourself for. How do you make it work for everyone? How can you manage people well when they're out of sight?

20. https://buffer.com/state-of-remote-work-2019

Priming Yourself for Supporting Remote Working

As a manager, you owe it to yourself to understand and support remote working because of either of these reasons:

- It's already happening in your company and team.
- Given the shift taking place in industry, it's likely you'll be working with distributed staff before you know it, even if it starts small with your team increasingly working from home.

We're going to touch upon the following topics:

- *Organizational inertia*, where you find your staff increasingly wanting to not be in the office, which is causing tension with how the company currently operates.
- *Good communication practices* that allow staff to feel included no matter their location.
- The importance of *trust* when managing distributed staff.
- Why *values and goals* have to trump informal interactions.
- *The importance of face-to-face time* and what that means for your travel budget.

Fighting Organizational Inertia

If your organization doesn't already support remote working, then it'll only be a matter of time before your staff begin to ask you about it. You may find yourself caught in the middle between requests from your staff and an organization that believes that the only way in which it can be productive is when everyone is co-located in the same office. This is simply not true: supporting flexible and remote working models is the future of work, and you need to create the environment within which your staff can get used to that way of working.

If your organization is resisting, then put your own reputation on the line as a manager. Try to pitch an experiment where each person in your team—should they want to—works one day at home every week. Tell your organization that you are responsible for their output. It's up to your team to decide which day of the week works for them best. For the true distributed working experience, see if everyone can work from home on the same day. If the company is still resisting, then be radically candid with them: to be unable to support remote working is a death knell. The industry is leaving them behind.

If you can proceed with this experiment, you'll see that it isn't so bad after all; it's likely that many of your staff, depending on their setup at home, will find that they can focus more when they're not sitting in a busy office. After a period of time—say a few months—do a retrospective on how the experiment is working out. What do people like about being remote? What do they dislike? What are some of the challenges of not being physically present? How could you overcome them as a team? Has it developed empathy for staff who are permanently remote?

Hopefully your organization will see that there's nothing to fear about people occasionally working remotely, and that they might even be able to solve some of their hiring problems if they look beyond a twenty-mile radius of their office! Who'd have thought it?

Communication Practices for Distributed Teams

If the primary forms of communication between workers physically sharing an office is *spoken* and *synchronous*, then the opposite is true for distributed teams. The primary forms of communication are:

- *Written*. The majority of communication occurs via emails, chat, direct messaging applications, and pull requests.
- *Asynchronous*. Staff may not necessarily be available to talk at a given time or even on overlapping time zones.

For particular individuals—notably extroverts—shifting to this mode of communication can be a challenge. Writing succinctly doesn't come naturally to everyone and can take practice. However, there are ways to begin practicing this within your team right away:

- Encourage your team to *work in the open* by regularly sharing what they've been doing to a team chat channel. This can be pointers to tickets, pull requests, or ideas that they've written up for distribution.
- Adopt a culture of *documenting thought processes*. For example, when working on a ticket, document the steps that you are going through via the comments so that others can see how you explored and solved the problem.
- Take the time to *write up decisions, priorities, and actions* for sharing to the team using written communication. Sharing these via email is archival and gives staff equal access to information regardless of their location. This is especially important when conversations happen informally and without the whole team present.

Using more written communication solves *some* of the issues with asynchronous communication, but not all of them. Two of the main challenges it brings are:

- Some distributed staff have *minimal overlap* between their time zones. This can be challenging when there are *dependencies* between two distributed staff, such as needing specialist knowledge or waiting for a prior piece of work to be completed before starting another.
- The feeling of *anxiety* when you or one of your staff get blocked and need help from others who are not immediately available. You may also feel this anxiety *as a manager* because it's harder for you to see what others are up to at any given time.

The solution to these challenges, other than it becoming more bearable with time and experience, is to encourage all staff to take *complete ownership of unblocking themselves*, thus not requiring you to be the center of all communication, and also to have *multiple items that they can work on at a given time.* This way if they become blocked, they can simply pause that piece of work and start on another. Again, this may be uncomfortable at first if you or your staff have been used to being able to reach the rest of the team immediately by swiveling around on your chair.

Trust Is Essential for Remote Working

Organizations that expect bodies in chairs at a fixed time every day will struggle the most with the shift to remote working. Working with staff in other parts of the world who do a majority of their work while you're asleep requires a great deal of trust. You must resist the urge to constantly prod and poke those who are out of sight and let them get on with their work. That's meddling and micromanagement.

You need to shift your attitude to be positive. You haven't heard from Alice in six hours since stand-up? That's good, as she's probably focused on her work. Check in with her after stand-up tomorrow. As a manager, before you panic over getting an update to satiate your own needs, put yourself in their shoes and consider what it's like to receive what you're about to send them. Also consider why the tools that you're using are unable to satiate your desire to check in. Can the software that you use for tracking work be better so that you always have a stream of up-to-date information on hand if you need it?

Trust enables your staff to do their best work and be bold in their choices and actions. Don't undermine it because remote staff aren't as visible to you as the person sitting next to you. It all starts with you: be positive, trust them, and encourage autonomy.

Values and Goals as Guidance

With fewer opportunities for informal interactions, formal values and goals for your staff become more important.

- *Values* define the way in which the team works so that people can asynchronously make progress in a unified way. For example, is it important to bias toward action rather than asking permission, or is consensus always required? Should you move fast and break things or tread carefully as a team?
- *Goals* define what the team are aiming for so that everyone is aligned. With a distributed team, goals become more important since they allow staff to work autonomously. As a manager you should be working with your team to set goals collectively and individually. Then you can refer to them when planning iterations of work and also in your one-to-ones with your staff. Where possible, link goals back to measurable company metrics: are you making the company more money, or increasing the time that users spend in your application?

Defining Team Values

You can run a workshop to define team values. By taking half a day, for example, your team can learn about each other and suggest and discuss values that you feel are important. For each, you can then define:[21]

- What does this value mean to us?
- What does it look like in action?
- How might it be misinterpreted?
- How will we evaluate adherence to it?
- How will it change our relationships or our interactions?

These can then be drafted, worked on further, and ratified so they can further help remote workers understand how to act independently and autonomously.

---

21. https://hbr.org/2018/04/how-to-establish-values-on-a-small-team

Working Toward Goals

Setting a clear *goal* for each iteration or milestone that your team is working toward can be a uniting force that enables autonomy. Be clear and understand how to measure it. Perhaps for this iteration your goal is to ship a new feature into production or to speed up data ingest by 50%. Having this defined allows remote staff to work more independently and creatively. You can choose how formal you want your goal-setting to be. It could be as simple as a short sentence that you define when planning iteration (*increase the number of users that login daily by 10%*), or it could be a formal framework such as OKRs (objectives and key results).[22]

Face Time

Even when teams are fully distributed and predominantly interacting through written communication, seeing and being near each other never loses its importance.

Video calls are essential for resolving conflicts and discussing open-ended issues. They're also one of the few ways in which you can truly feel connected to the human at the other end of the computer. If you're managing remote staff, one-to-ones should be done this way.

When participating in video calls with a mixture of office-based and remote participants, pay attention to the amount of opportunities that the remote participants have to join the discussion. Those sitting together in the meeting room will more naturally bounce off each other with no lag in the conversation. You can have one person act as a "spotter"[23] to ensure that the remote participants are getting enough space to speak by watching visual cues.

Additionally, if you're one of the office-based participants, before the meeting starts, ensure the room has a good AV setup, that the microphone and the camera are working, and that you're all sitting in a place where you can be seen. It's the many little things that send the message that you care about those that are on the other end of the line. Don't tap on the table or bounce your leg. You can't hear it, but they can.

Remember: as soon as you have regular remote workers in your company, you'll need to treat *all* meetings as if they have remote participants present.

---

22. https://en.wikipedia.org/wiki/OKR
23. https://randsinrepose.com/archives/a-distributed-meeting-primer/

Video Calls: Do's and Don'ts

A few small tweaks can make the video-call experience dramatically better for all participants.

Do:

- Normalize the experience as much as you can. If you're sharing a video camera in a big room, then it's a better experience for remote staff if you're all partaking individually via laptops.
- If you have to use a shared webcam in a meeting room, arrive early to set up and test the equipment so that remote participants can see and hear you properly.
- Use collaborative drawing software or shared documents where possible, rather than pointing a webcam at a whiteboard. If you have to use a whiteboard, ensure you write and draw in a way that makes it clear for those that are remote.
- Occasionally experience what it's like to be a remote participant by being one yourself. Switch up days working from home so that you have to be remote in that group meeting. How does it feel?

Don't:

- Ignore remote participants. Have one person act as a spotter to bring them into the conversation and watch for physical cues that they want to talk.
- Talk over each other. It's hard to pick apart the noise and also reduces the spaces in which remote participants can contribute.
- Fiddle with pens, tap your fingers or feet, or doodle aimlessly near the microphone. Those small, unnoticeable sounds on your end are distracting and annoying on their end.
- End an important meeting and forget to follow up on the main points that were discussed. Writing a few bullet points to participants ensures opportunities to increase understanding and to ask questions.

The other part of face time is *real* face time. Where possible, get any remote workers on your team into the same physical location as the rest of the team a few times a year. Yes, the travel is expensive, but it makes all the difference. Meeting people in person adds a whole additional human dimension to your relationship, diffuses months of tension in an instant, and builds trust and rapport. It's well worth it. If your department does quarterly or annual meetings, fly them over to be there in person.

Work-Life Balance

As technology has eaten the world, many companies that have created that technology have also eaten their people. We're all influenced by the practices and celebrated successes of the tech giants that we see publicized in our industry, and those celebrated successes are the "unicorns," the 10x rebels—hustling hard, being disruptive, and vacuuming talent and competitors in a race to become the dominant force in the market.

We read books, we watch podcasts, we follow influential business people on Twitter, and slowly, with time, we move the norm toward the extreme with the amount that we are all expected to work. The technology industry is obsessed with growth. Valuations of companies are typically driven by their revenue and their year-on-year growth percentage. VCs push founders and their boards to deliver significant multiples of their initial investment.

2x isn't enough. 5x is middling to average. 10x is the number that has become a cliché: 10x thinking, 10x scaling, and 10x engineers. You may well work for one of these companies. You may be looking up at the wall of your office and seeing a slogan like "*Don't stop when you're tired. Stop when you're done.*"

Within this culture of hustle is a clear and present danger: the promotion of burnout and workaholism to people that don't have the experience to realize that it isn't going to end well for them. This is where you can come in as a manager. You can be a counter-balance to a disease that has become endemic.

Protect Your Team

As a manager, you can set a strong cultural precedent by acting against a foolish culture of harder and faster at all costs. Instead, establish your own culture.

This should be where your team:

- Work a *sensible number of hours* a week that allows for people to have families, dependents, children, and hobbies outside of work. Your staff should be mindful that work is only one part of their colleagues' lives.
- Don't rely on messages out of hours to get things done. Move to an asynchronous communication pattern.
- Celebrate the work that they do, whether it's a success (you provided value for your users) or a failure (you learned of another path that wasn't worth going down).
- Remain focused and efficient and able to achieve their goals by saying *no* to anything that is unimportant. This is a key area where you can help.

By adhering consistently to these rules and still being able to deliver high-quality work, you can begin to inspire a cultural shift away from being busy idiots and toward being focused professionals that do great work but also have balanced lives. Work on increasing the effectiveness of the existing work hours, rather than adding more.

Walking the Walk

Of course, none of this matters if you don't do the same yourself. Don't be a manager that encourages the previous practices but then still sends emails to your team at 11 p.m. or stays in the office until 7 p.m. You need to walk the walk as well as talking the talk. Otherwise you're setting the wrong cultural precedent. You need to be aligned in your words and your actions.

One workplace initiative that works well is to *leave loudly.*[24] It encourages leaders and managers in all areas of the business to combat *presenteeism*, which is where people sit around at their desks waiting for others to leave before they leave, often doing little that is productive. Instead, leaving loudly encourages people to announce that they're signing off for the day: "I'm off to do the school run," or "I'm outta here to get the groceries in," or even "I've finished much earlier than I thought today. Time to head home."

By walking the walk, you leverage your influence as a manager to show that if it's OK for you, it's OK for everyone else, too, and that's the way that it should be.

24. https://www.news.com.au/finance/work/leaders/why-pepsico-ceo-asks-his-team-to-leave-loudly/news-story/5467b3ffff387c3a5dd79ac3a245c868

Notes on Culture

The term *culture* has been used many times in this book with little explanation. Sorry about that. Entire books are written on what company culture is and how you can design your own, so the full breadth of the subject is somewhat out of the scope of these pages. We recommend *What You Do Is Who You Are: How to Create Your Business Culture [Hor19]* as a place to start.

But in short, culture is the effect that your company, your team, and your own management style has on others. It has nothing to do with having a ping-pong table or beanbags or an iced-coffee dispenser. It's about the psychologically safe environment that your staff enjoy doing their work in.

Regardless of the *company* culture, the culture that you set within your team comes through your *actions*, intentional or unintentional. That's why this book has given you a framework in which you can practice being a manager: how to do everything from handling your emails and to-dos, through to hiring and dealing with performance issues, through to the nuances of how to exist within the wider workplace. This ensures that you're able to contribute toward a professional culture by being effective, mindful, and honest.

However, one important ingredient needs to be added to the recipes that you've read so far: *you*. For a culture to be successful, you have to be yourself. Many managers, upon being promoted, start becoming someone other than themselves to appear authoritative and in control. Don't do that. Just be yourself. It doesn't matter what you've read about other successful managers in other companies or how many talks you've seen elsewhere about how it should be done, because *you will create your own team culture by being yourself while following the advice in this book*. You don't need to be anyone else. This means that you may have cultural differences with how your own manager works, which means you need to work harder on that relationship.

You'll always have parts of you that need some work. For example, you might need to work on your written communication or on keeping your cool in frustrating situations. However, these are weaknesses that you should find by being yourself, and then you get to apply those fixes *to yourself* as well. This is why management can be such a rewarding profession: not only do you get to work on improving a team, you get to work on improving *yourself*. Not bad for a day in the office.

Your team culture will come from you. Be yourself fully and do a good job. The rest will follow.

To the Land of Unicorns

In this chapter you've learned about some of the challenges facing our industry and what you, as a manager, can do about them.

Here's what we've covered:

- The pressing issue of *diversity* in the technology industry. We've looked at what it is, why it's critically important, and what you can do to begin tackling the problem.
- *Flexible and remote working*, which we'll only see more of in the future. We've learned some ways in which you can better support remote workers and be ready for the shift that is already underway.
- The importance of *work-life balance*, or in short, how to not let work take over your or your team's lives. As a manager and leader you need to enforce this via your own *actions* rather than by rules.
- Lastly, we touched upon *culture*. Fundamentally, you can make culture better by being yourself and doing an excellent job. It really is that simple. But not easy.

All of the subjects that we covered in this chapter have many excellent books written about them if you wish to dive deeper into any of them. Start with those that have been referenced and see where you go from there.

In the next chapter we're going to be looking at the world of startups and considering how your newfound managerial skills could make an impact while accelerating your career.

Game's the same, just got more fierce.

Slim Charles

CHAPTER 17

Startups

You open the link that Paulo just sent to you.

```
Local startup makes it big with $3-billion valuation
```

"No way! Is that the office that's just around the corner?"

"Yep. That's them," replies Paulo.

You scan through the article. "They only started two years ago? And they're making that silly application for arranging deliveries?"

"Look at the quote at the bottom," says Tara.

"No way! It's Sam who used to work with us!"

Paulo rolls his chair over and points at your screen. "Look at her job title."

```
...notes Sam Harris, VP Engineering.
```

Your shoulders tense up. "What! How on earth is she a VP?"

"I know, right?" replies Paulo.

Tara has a tone of jealousy. "And now she's probably going to be filthy rich as well. Imagine what her stock options are going to be worth!"

As you reach the end of the article, you experience a mixture of anger, envy, and regret. You interviewed at that small startup before accepting the offer for your current job. You went with the bigger company and the higher salary.

You look at Sam's job title in the article once more. *VP Engineering*. You remember the day that she told you that she was leaving. Little did you know that she was leaving to leapfrog you in your career by about ten years.

As you turn your attention to the busy city outside of your window, the questions begin to arise. Are you working for the wrong company? Should you be getting in early at a startup instead? How did Sam manage to make VP? Could *you* have had that fast career growth, being able to buy that house that you always wanted?

This chapter is all about startups, and we've got many questions to answer. What are they? How do they usually form and grow? Should you be considering leaving your current company to take a lower salary for a bigger responsibility elsewhere, or is the risk too great? And are managers even needed among the hacking and pivoting?

Here's what we'll explore:

- We'll touch upon the *concept and life cycle* of startups, and why they've generated so much attention and interest in the technology industry.
- Then we'll work out what kinds of *opportunities* exist for you as a manager if you were to work for a startup.
- Lastly, we'll debunk opinions that management is something that only big companies do. Instead, we'll create a case that the skills that you've been learning in this book are critically important in early-stage companies.

Software Is Eating the World

Unless you've managed to avoid media coverage for the last decade, you're aware that technology companies are often making the headlines. As Marc Andreessen wrote back in 2011,[1] software really is eating the world. Internet companies like Facebook, Twitter, Netflix, and Google have become part of the fabric of everyday life.

We've seen habits that are centuries old change overnight with the introduction of taxi-hailing applications such as Uber and Lyft and food delivery applications such as DoorDash, and it's becoming more likely that you'll meet your future partner by swiping right rather than approaching strangers in public.

This ability for companies that offer their services via software and the internet to become worth more than Walmart in less than 1% of the time has attracted a lot of attention. It has highlighted the possibility that a good idea, combined

1. https://a16z.com/2011/08/20/why-software-is-eating-the-world/

with investment from wealthy venture capitalists (VCs), can potentially create magical *unicorns*: another word for a company that ends up being valued at over $1 billion.

For those startups that find their niche, grow rapidly, and then exit via acquisition or listing on the stock market, early-stage employees with shares in the company can make millions. In some cases, billions. Smart engineers worldwide are drawn not only to new and innovative—often society-scale—problems to solve but also to the ability to get in at the ground floor, experience career progression that grows as fast as the company, and have a lot of say in how things are done.

In addition to the potential reward, the fast pace, and the cutting-edge computer science, startups provide many other attractive benefits. Since they are young companies, they're often run by smart, decisive, and often rebellious founders wanting to do things *their way*. This means a focus on fun and enticing office environments, plenty of employee perks and benefits, and company cultures that celebrate autonomy, meritocracy, and the ability to get impactful work done, rather than being the best politician.

You may have been tempted to apply for a position at a startup for one or many of the reasons mentioned above. But the question is: should you? After all, you've gotten this far in the book because you're invested in honing your craft as a manager. One could argue that increasing your impact as a manager requires managing an increasing amount of people, and startups don't have many people at all. Additionally, many startups are focused on finding their product-market fit rather than establishing their internal hierarchy, so without that, what good is management?

Let's begin by looking at the general composition of early-stage startups and then look at various scenarios in which your managerial skills are in high demand.

Opportunities for Managers

Startups move through numerous phases from their inception as an idea through to their growth as a full-fledged company. As these stages progress, various opportunities can arise where you can get involved as a manager. We'll take a look at these with reference to the diagram on page 330.[2]

2. https://en.wikipedia.org/wiki/Venture_capital

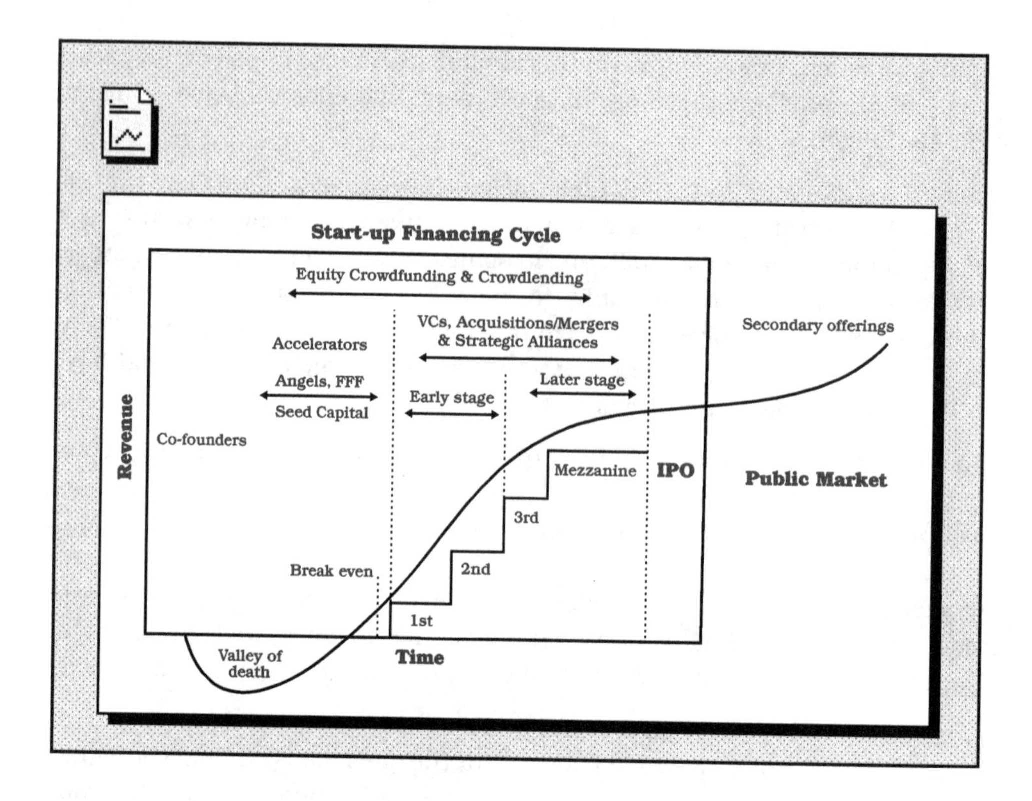

Let's begin by briefly walking through the diagram from left to right, taking in the stages.

- *Co-founders*: This is where the people with the initial idea create a business and get started. Unless you're one of the founders yourself, you won't be concerned with this stage. The co-founders will be looking for a *problem-solution* fit: is it possible to build a product that solves some business problem? Typically this results in building a minimum viable product (MVP) to see if customers purchase it and what they can learn from them. Most, if not all, input at this stage will be from the co-founders.

- *Initial investment*: If the initial MVP gains traction and looks like it may become a viable business, then startups seek initial investment so they can begin to grow faster than they could organically. This initial investment can come from angel investors (typically wealthy businesspeople), startup accelerators, specialized investment funds, and more. It's enough money to see whether an increased investment in the product can increase its impact in the market. A big chance here is to get involved as *manager number one*. We'll talk about that shortly.

- *VC rounds*: Assuming a startup's product continues to do well, then the company may decide to take on even more investment to keep growing faster and faster. VCs are private equity investors that grow their funds by investing in fast-growing businesses. During these stages a *number of opportunities* can arise as that money allows many more engineers to be hired, teams to be created, and possibly even create positions that allow you to run multiple teams.
- *Beyond*: Once a company grows beyond the VC stages, it's probably beginning to get as big as many other big companies!

We'll look at the two previously mentioned startup stages that bring a lot of opportunity: the stage that brings initial seed investment and the VC rounds that will follow.

Becoming Manager Number One

One of the biggest career growth opportunities for managers is becoming the *first* manager. This usually happens after some initial investment. But what does that mean? Typically, within the co-founder group, there will be a CTO and a CEO.

The CTO, despite the big job title, will typically be the lead engineer that is building the technology. Their priority will be getting the product built so that the MVP that has gained traction can continue to scale, support new features, and serve existing customers. The CEO will be usually focusing on all manner of things: new sales, hiring, building a plan for the future, and so on.

What this means is that the engineering department doesn't have anybody focused on management. By management we don't just mean people management either: we mean the management of process, hiring, communication, and delivery—the person that keeps the wheels turning efficiently. This could be you.

You would be expected to do what you're doing as a manager of a team, but the scope of that role covers the entire department, even if it's small. The impact is therefore felt across the whole company. The seeds that you can begin to plant at this stage—assuming that the startup does well—can blossom into the engineering culture of a department of tens or maybe hundreds of people.

And what is this role called? It's typically given the title of *VP engineering*. If the CTO is creating the product, then the VP engineering is responsible for creating the engineering organization. Here are some of the things that a startup VP engineering may be responsible for:

- *Ownership* of the delivery process, from technical aspects, such as how to deploy frequently with no downtime, through to the way in which releases are communicated to users and to the rest of the company.
- *Performance* of the engineers, encompassing line management, one-to-ones, performance reviews, and creation of all of the processes that surround those activities.
- *Resourcing and prioritization* of projects. What's going to get done and in which order, with limited engineers available?
- *Hiring* new engineers, including the creation of the hiring process in the first place, and working out which roles and skill sets are required.
- *Keeping the CTO focused on building.* If there's anything that doesn't fall into the examples above that is getting in the way of the CTO, then it's likely to fall on the VP engineering to solve it.
- *Building stuff!* Given that the company is so small, it's unlikely that a startup VP engineering isn't contributing code.

When you look at the VP engineering role from this perspective, you could imagine it as an engineering manager role with the accountability, risk, and autonomy turned up to 11. If you don't know how to do something, or if there's no process at all for some particular issue, then guess what? It's up to you to solve it!

The relationship between the CTO and VP engineering at a startup can be summarized in the matrix diagram on page 333. A good VP engineering can turn a small band of pirates into an efficient naval vessel, with plenty of room for the fleet to continue expanding.[3]

So, remember that this early stage of life for startups could offer an exciting career opportunity for you. Don't be put off by the lofty job title. Instead, if you continue to enjoy and succeed in management, and if you're feeling up for a challenge and the ability to have a large impact, then a startup that has received some initial investment may be looking for someone exactly like you. Just keep an eye on job advertisements that offer the role where the team size is small, often in the low tens of people, and lean in to your network to find out what's not yet advertised. Don't be put off from applying.

---

3. https://bothsidesofthetable.com/want-to-know-the-difference-between-a-cto-and-a-vp-engineering-4fc3750c596b#.gw

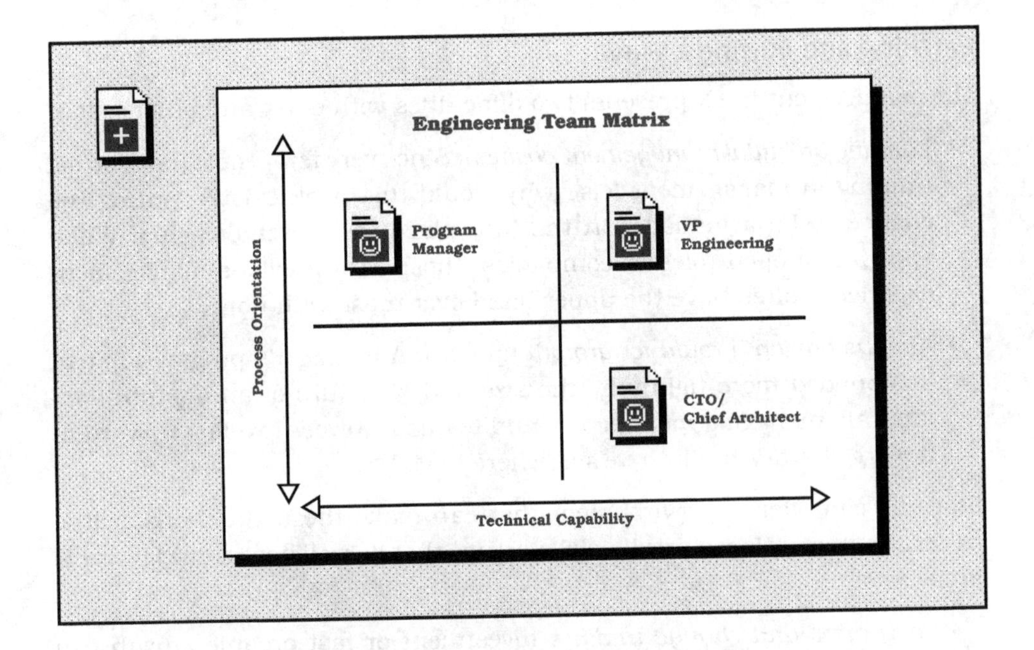

In addition to the challenge and excitement that this role can bring, if the startup continues to succeed, you may have the opportunity to define and hire a substantial department, giving you the opportunity to move into a manager-of-managers role much quicker than you'd be able to at a larger organization. It's the perfect opportunity to tee up much bigger roles in the future.

So yes, *you* could be VP engineering. Maybe even sooner than you think.

Your Turn: Could You Be Manager Number One?

Consider your own career. Although you may be new to management, could you imagine putting yourself forward for this startup VP engineering role in the future? If so, why does it interest you so much? If not, what do you think may be preventing you from wanting to do it?

When working at a startup, you'll find things are often challenging because of the high degree of ownership and accountability. You'll encounter twists and turns, failure and success, all multiplied by 100. This means that you had better be passionate about what the startup is building! What sort of technology would you need to be creating to make you want to do this role?

Catching and Surfing a Wave

The management track presents two difficulties with entry and progression:

- *Landing an initial management position.* Since very few people have formal training in management (else why would this book exist?), getting the chance to be a manager in the first place can be challenging. When applying to open roles at companies, those with previous management experience often have the upper hand over those with none.
- *Always having a route for growth upward.* A manager's progression can be bounded more tightly by the size and structure of an organization. After all, managing more people and teams involves traveling upward in the org chart. What if there's nowhere to go?

Startups can offer an environment that can make these situations a little easier. As we mentioned previously, startups that do well have some favorable properties:

- *They grow and change rapidly.* Investment or fast organic growth can double or triple the size of the engineering team in a short space of time. This means that a handful of people working in a flat hierarchy may now require more structure in their org chart, including the introduction of more line management. This creates roles you can step into, especially if you're already working there. Additionally, continued success can expand the department further, and you're well placed to continue the journey upward.
- *They have less money to spend.* Given that startups are unlikely to be poaching highly paid managers at larger technology companies, you can use this to your advantage: you'll have more of a chance to make the jump into management even if you haven't had prior experience.
- *Each individual can have a big input in decisions.* If you join a startup as an individual contributor with a desire to move into management, then you'll have a lot of say in creating the conditions in which that can happen. Your influence at a small company will be much greater than that of a large corporation. You can make it happen for yourself.

Sampling Different Roles

Another benefit of joining a startup is being able to sample all manner of different roles. Given that there are limited people, it may be the case that you get to lean into—or even try your hand at—various skills that you may not have at a larger company.

Sampling Different Roles

Perhaps you get to be a full-stack engineer rather than just a back-end one. Perhaps you may get to contribute toward product design or the roadmap itself. There are many opportunities for self-starting individuals to have fun and round out their resumes.

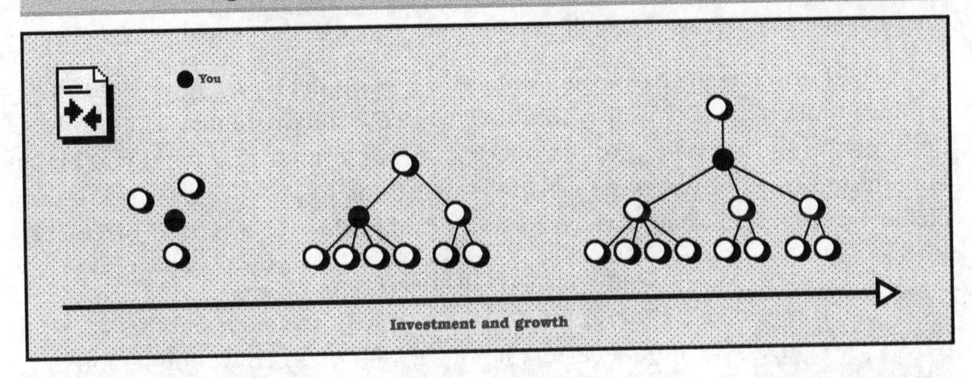

So here's some advice for how you could approach and join smaller companies in a way that allows you to have your career as a manager in mind:

- *Co-founder stage*: If you know of any startups you're interested in that are just getting going—perhaps you know them through colleagues or friends or have met them at a conference or meetup—then *register your interest*. Tell them that what they're working on seems exciting and that if they begin to grow that you'd love to get involved, potentially as an engineer with scope to grow into a manager. Have these people as part of your network.

- *Initial investment stage*: When startups raise money, they often put out press releases. Make sure you stay on top of what's happening in the industry by reading websites such as TechCrunch, Hacker News, AngelList, and Product Hunt. Companies getting initial raises are still small. Find out who works there and reach out on LinkedIn or Twitter. Be direct: say you're looking to either land a first management role or grow as a manager, and you're interested in what they're building. See if you can make connections. If you're already experienced, are they looking for manager number one? You never know what'll happen.

- *Subsequent VC rounds*: As startups grow to a size where they begin to raise VC capital (often called Series A, B, C, or beyond depending on how many prior raises there have been), then you're likely to be dealing with a more standard recruitment process. However, a short cover letter stating your intention to go down the management track in the future, even if there are only individual contributor roles advertised, can open other doors for you. Just state your intention.

A world of exciting small companies is out there. However, there are risks. Joining a startup may be intense, difficult work, without the structure, compensation, and benefits that you've been used to in the past. The environment may not be for everyone. You'll have to work out whether it's right for you.

Not only that, most startups fail. Many have great ideas that never find their product-market fit. Some are unable to raise enough money to continue. Perhaps the startup may pivot its offering so many times that what you build is nothing like what you were originally expecting to be working on. Are you comfortable with change? Are you a cathedral builder or a bazaar browser?

Startup Compensation and Equity

It is unlikely that startups will be able to pay as much in base salary as established technology companies. They simply don't have as much money available at early stages.

You may find, however, that they offer equity in the company as part of the compensation package, typically in the form of stock options that vest over some period of time.

Before you get sucked into dreams of becoming a millionaire as the NASDAQ bell rings, you need to remember a few things:

- Most startups fail, and even those that are successful may take over ten years for the equity to be exchanged for real money. Don't count on it ever happening. Be there for the experience instead. Everything else is a bonus.
- If you're unsure about the terms and conditions of any offered equity, including how much there is, whether it's a preferred stock class, or anything to do with the vesting schedule, then seek legal advice.
- Betting your happiness on an uncertain and unpredictable future will likely lead to disaster, even if that disaster is confined to the boundaries of your own brain.

It's not all gloom and doom though. Due to their size, startups expose you to all aspects of a business—from engineering, to sales, to marketing, to angry clients—often because the lack of people means that everyone needs to support each other closely, regardless of their role.

If you're interested in expanding your knowledge of how a whole business works, they're a fantastic way of being able to get up close and personal in a short space of time. One year at a startup can feel like five years at a larger company because of the intensity and variety of the work. You can move on to your next role as a more fully rounded professional.

Remember that startup experience is highly sought after because being impactful in that environment involves being enterprising, self-motivated, collaborative, and quick to learn. Even if the startup itself doesn't work out, your next gig will be all the better for it.

Your Turn: What's Out There Right Now?

Have a look at the websites mentioned in the previous section, like AngelList. What kinds of startups are out there right now that look interesting to you? Are any of them hiring formally? Do you feel brave enough to send them an email saying hello?

Are any of them hiring a VP engineering, or similarly titled manager number one? What does the job description look like, and does it look like something that you could do if you increased your experience as a manager? Is this exciting to you, or too risky?

Why Management at Startups Is Critical

At this point, you should see that there's plenty of scope for using startups as an opportunity to land that first management position, or perhaps, after you've become comfortable and tenured with management, making a giant leap into being VP engineering.

However, as you're probably aware, many engineers have a stereotypical view of managers. Perhaps they feel that management is busywork when compared to the "real" work of getting the product out the door. Perhaps they think that it's where the bad programmers go to get a pay raise. Perhaps they think that it's entirely pointless altogether.

I hope that if you had any of these feelings when you started the book, that they've now dissipated. However, there may be people out there that think that management has no place in startups and small companies. They may feel that management is something for companies with 1,000 employees rather than 10.

After all, some startups take more radical approaches to management such as self-organization via Holacracy,[4] where the organization is designed not as a typical tree structure—where reporting lines cascade down from the CEO—but as self-organizing *circles* with clear purpose and accountability. Other alternative governance structures include *Teal [Lal14]*, which promotes self-management guided by consistent practices, purpose, and values.

Yet, regardless of how the org chart is constructed, and regardless of the stage of the company, the skills that you've been learning in this book are highly applicable, and often more widely than you think. Even if you aren't managing a team, you can benefit from better self-management, a mindful approach to communication with individuals and whole departments, knowledge of how to advertise for roles and how to interview candidates, and so on. You can apply what you learned about the challenge of working with humans and on complex projects *anytime*. Management skills transcend the role itself.

Additionally, management, even at small companies, provides employees with a scaffold for their growth and personal development. Management doesn't mean bureaucracy. Good management is a light touch and continued collaboration. This doesn't get in the way of anything. Being an excellent manager at a startup breaks the stereotypical norm. What if employees not only got to work on exciting and innovative projects that could hit the big time but also got unrivaled support and career development opportunities while they were there? Sounds like a win-win situation to me.

Those who are able to adopt a management *mindset* can provide a much-needed counter-balance to the speed at which startups can grow and become chaotic. Whereas the company and the CTO may want to press forward with the pedal to the metal, those with management mindsets can do so while keeping an eye on process, efficiency, team organization, communication methods, quality, and organizational scalability. Management skills can maintain equilibrium. And if you're able to do this within a small team, you might find yourself rising up through the ranks faster than you think. I think you'll agree that the future is bright.

What Does Your Future Hold?

This brings us to the end of our discussion around startups. Hopefully you now can see that they can offer you a number of opportunities if you're willing.

4. https://www.holacracy.org/

More importantly, I believe we've shown that the *practice* of management transcends large companies and complex org charts. You can make an impact *anywhere*.

Here's what we've covered in this chapter:

- We looked at the *concept and life cycle* of startups, and why they're a compelling prospect for exciting work.
- Then we touched on *opportunities* for you to get involved, whether that's using them as an opportunity to accelerate your career progression or perhaps even joining as VP engineering.
- We closed out by providing a *counter-argument* to those that may think that management isn't needed at startups. Management is not indicative of stuffy bureaucracy. It's a mindset and skill set that can provide a much-needed counter-balance to chaos.

With all of this talk of startups, we've possibly made you think about your current role and where you want to progress in the future. Perhaps you already have a clear idea of where you're going, or perhaps you haven't thought about it at all.

In the next—and final—chapter of the book, we're going to go through an exercise together that maps out your career universe. Then we're going to look at setting some goals for yourself over the next few years. The bonus is that you can do this with all of your staff too.

Steer the course, make a way, and come ashore on a greater day.

Yasiin Bey

CHAPTER 18

The Crystal Ball

You stand by the kitchen window with your coffee in your hand. It's 7 a.m. and the rising sun envelops the surrounding buildings in an orange glow. You cast your eyes across the worktop. From left to right: bread knife, coffee grounds, ID card.

As daybreak begins to warm the room, you remember looking down to see that card in your hand on your first day at your current job. It's true: the days are long, but the years are fast. One has just passed by. Winter is making way for spring. The old becomes new. Buds unfurl and gloves go back in the cupboard.

You had a new start a year ago. It started with trepidation. A big career move, a leap into the unknown. On reflection, things turned out fine. Your team is talented and you feel like on most days you've been doing a good job. Mostly. You've made new friends and connections. Like Indiana Jones, you chose *wisely*.

But what does the future hold? You think about why you took this job in the first place. It was for a multitude of reasons: a new city to explore, a new company to immerse yourself in, and a new role to master. However, your career still has plenty of years left to come—decades, in fact. In some sense, you're only getting started. When you're standing with your morning coffee five years from now, where will you be? What will you be doing? Most importantly, will you be happy?

These are all fundamental questions that have complex answers. After all, the future is never fixed. It's like looking out of the window of a moving train. It manifests through a mixture of luck, opportunity, willpower, and acceptance.

This is the final chapter of the book. Like most endings, it's actually a new beginning in disguise. With all of the tools in place, you now have the ability to imagine where they could take you. We're going to be doing an exercise together that allows you to think about your career development, although to some extent, your thoughts will be broader than that, covering life, the universe, and everything in between.

Here's what we're going to look at:

- We'll *think about the future*, and why it is so important for you and your staff to continually revisit where they want to go and who they want to be.
- Then you'll go through an exercise that defines your career *vision*. It'll be fun. We'll think about the past, the present, and the future. We'll dream a bit. Then we'll get into the details of how you get there in your *plan*.
- We'll wrap up the book by condensing the exercise into a concise set of instructions so that you can *do the same with your own staff*.

Grab a pen and paper for this chapter. It's interactive. Make some time, make a coffee, and get comfortable. We are going to travel through time together.

Life, the Universe, and Everything in Between

Human beings, for as long as they've been on this planet, have pondered the meaning of life. Why are we here? What should we be doing? What does it all mean? The answer to that question depends on the individual. Some find inspiration in their faith, others in their families. Some seek power and wealth, whereas some seek simplicity and peace. Some believe that the answer is 42.

Research into the areas of the world where people live longer and healthier lives [Bue10] identifies a number of common characteristics exhibited by those that live there. In addition to somewhat expected lifestyle traits such as moderate, regular physical activity; a plant-based diet; and engagement with family, close friends, and the local community, an important part of a long and fulfilling life is *purpose*.

Purpose is not about economic status or a feeling of well-being. It's about a life worth living. Okinawa Prefecture in Japan is one of the areas identified as a hotspot for human longevity. The term *ikigai*, which roughly translates into English as *what you live for*, is thought to have originated there. *Ikigai* is about identifying the source of value in one's life: having a compelling desire or goal for the future that propels you forward.[1]

Work is where most of us spend a significant portion of our lives and is also where we focus a significant portion of our attention. You're reading a book about management, after all! For those of us fortunate enough to have skills that are broadly applicable, in demand, and compensated well, we often have the ability to exercise a great deal of choice in exactly what we want to do with our careers: where to work, what to assist in creating, and how we wish to develop as professionals. While this can be seen as a luxury, it can also be seen as a curse: the overwhelming amount of choice can lead us into a state of paralysis and can breed a continual fear of missing out on other opportunities.

Some of the communities that exhibit the greatest longevity and happiness, ironically, are those in which socioeconomic and geographical factors somewhat *limit* their choice in how they live their lives. They eat healthily because nuts, fruit, and vegetables are often the cheapest food. They exercise more because they have to walk more in their local area. Their choice of work is limited, so they're more accepting of what they have. Sometimes having fewer options works out better.

That's why the exercise that we're going to do in this chapter is so valuable for you and your staff. Not only does it give an opportunity to be introspective in a structured and supported way, it allows those that take part to consciously study and therefore *narrow down* the direction in which they are going in the short term by defining a compelling future for the long term.

An important thing to remember about our careers—and our futures—is that they are fluid. As we progress through our lives, our vision of the future will change. What did you think you were going to be doing with your life when you were eleven? How did that compare to what you thought at eighteen? What about now? It's guaranteed to change in the future, too. Even within the world of work, nothing is permanent. In the United States in 2018, the

1. https://ikigaitribe.com/ikigai/ikigai-misunderstood/

average length of tenure at a technology company was recorded at around four years.[2]

People move on all the time to seek new opportunity, higher salary, novel challenges, and varied locations. Given that lifetime employment at a company is a thing of the past, we need to accept that our staff will eventually move on somewhere. However, while they're here, we should look to define *tours of duty [HCY14]* that allow them to reach the next step in their career without operating under the false pretense that they're going to keep them forever. The same applies to you.

This means that we need to keep two things in mind at any given time with respect to our careers:

- The current *plan*, which is usually defined in a period of a year.
- The *long-term vision*, which we're always moving toward.

That's what we're going to do in the following exercise. We'll begin by thinking about the future. We're going to go through the exercise together first, and then once you've done it, we'll summarize how you can deliver it to your staff via a series of coaching sessions.

Your Vision

In an ideal world, every job that you have should be a meaningful step toward the vision that you have for yourself in the future. Where you work right now has the opportunity to be a career-*defining* place to work as long as your short-term and long-term trajectories are in alignment.

We'll begin by focusing on this vision. This is where so many people struggle, often for two reasons:

- They don't dream *big enough*, and severely underestimate what the future could hold for them.

- They aren't *clear enough* about the vision so that it has a compelling, magnetic force that propels them toward it.

So, that's the place where we're going to start.

How Did You Get Here?

No matter where you are right now, the journey you took—in hindsight—probably wasn't what you would have predicted. But first: did you make that coffee?

2. https://www.thebalancecareers.com/job-tenure-and-the-myth-of-job-hopping-2071302

If not, grab one. Get your pen and paper and draw a line similar to the one in the following diagram:

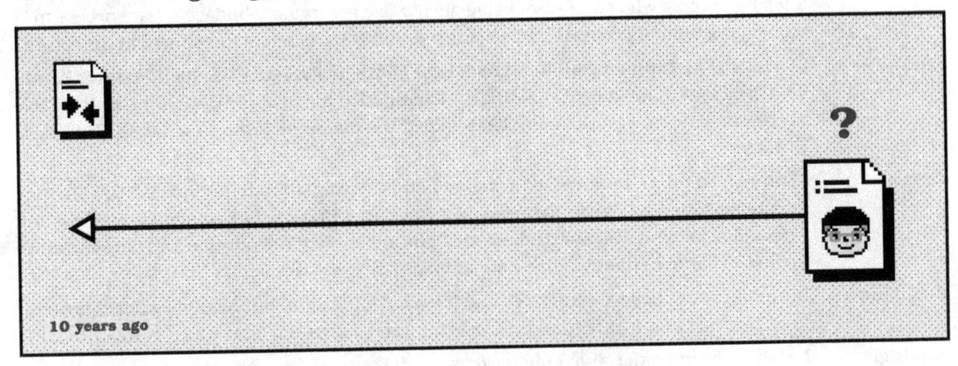

This line represents the last ten years of your life. Plot significant events in your lifetime against this line. Don't limit yourself to events in your career, such as landing a new job or getting a promotion. Instead, expand the boundaries out to touch the edges of your life. Include education, moving location, and significant events: get it all in there. Once you've done it, take a step back and let it all sink in.

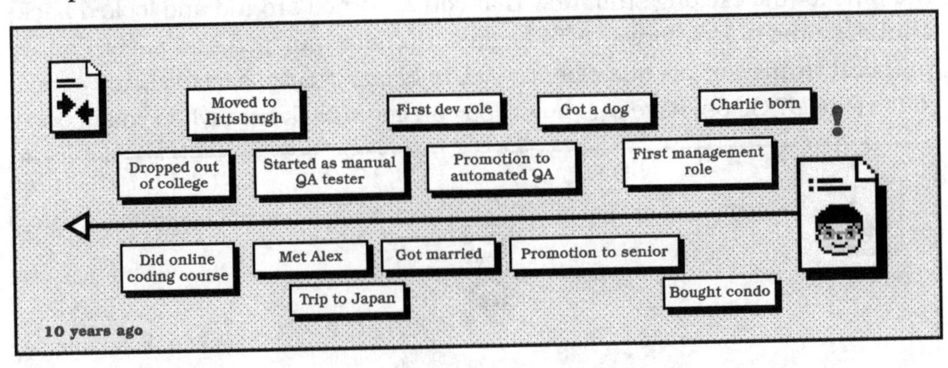

Many people significantly underestimate the number of things that can happen in a ten-year period. How do you feel about your achievements now that you're looking back at them on your own timeline? Would you have ever predicted that they would have happened?

Reflect on what you've learned both professionally and personally in this time period. Consider *training, luck, people, forethought, and hindsight.* Which role did each of them play in the unfolding of events on that timeline? Make some notes for each one. Perhaps you could arrange this like the table on page 346.

Area	How It Played a Part
Training	Teaching myself to code wasn't formal training, but it paid off in dividends. I now manage a team of people who code! Also, being able to use my training budget to have access to online programming courses while doing manual QA was a real investment in the future.
Luck	Moving to Pittsburgh meant that I met Alex. That's lucky. I was also lucky to meet Sarah, who lived in our apartment building and persuaded me to apply for the engineering role at the company she worked for. That helped me make the jump from QA to Back-end Engineer.
People	John helped me so much when I was making the transition to automated QA. He was an amazing mentor. When I started as a Back-end Engineer I had similar mentorship from Mihaela, who was able to help me fight my impostor syndrome and help me ramp up fast. Charlie certainly helped me manage my time better...
Forethought	Dropping out of college was hard, and it was a gut feeling, but I knew I wanted a career in technology and I wanted to be a programmer. Turns out that was a great decision.
Hindsight	I wish I'd been bolder at asking for doors to be opened for me at the beginning of my career rather than waiting for others to open them. Once I became more proactive at seeking what I wanted, more opportunities opened up for me.

Did all of those factors play a part in your past, or are some more prevalent than others? Which do you think has been the most important for you?

What About the Next Ten Years?

OK, time for the exciting stuff. Now that you've turned around and looked backward, it's time to look forward. To begin with, take your pen and paper and try your best to answer this question: *what would you like the next ten years of your future to hold?* Go on, have a go now. You can come back here when you're done.

How was the exercise? Was it easy or hard? Do you like what you wrote down, or did you struggle to write anything at all? Does your vision give you goosebumps, or does it seem thoroughly underwhelming? Does it struggle to look past the next promotion or pay raise?

It's time to iterate on it again. This time, here's some more prompting to get you thinking. Then, follow the instructions in *Your Turn: Detailing Your Vision.*

- How do you feel about your current job? What do you enjoy the most? What do you wish you could change?

- How do your skills currently map to the most innovative work happening in industry, and are you traveling in that direction? Do you want to be?
- Do you ever experience days in which you're happy to jump out of bed and head into work? If so, what drives that?
- Conversely, are there mornings that you can't bear lifting the covers? Why?
- If you haven't already, try the following: imagine yourself waking up in the morning, ten years in the future. What does your bed look like? What about the bedroom and outside the window? You head to the kitchen to make breakfast. Visualize where you live: the walls, the floors, the view outside, the people—if any—that surround you. You get dressed and look in the mirror. What do you see? What are you wearing? Where are you going, and how are you getting there? Dive deep into future you. Embody that person; really *feel* what it is to be them. After all, it's going to be you.

Now that you've thought through these questions, has that changed what you wanted to write in your vision for the next ten years? If so, go and do that. Is it a fair summary? Now it's time to construct your vision in more detail.

Your Turn: Detailing Your Vision

Let's build your statement out a little more formally. In a document, write the following:

- Your ten-year vision. Use compelling language to make it seem almost touchable. Try to use "I" statements: "I will be the CTO of a Fortune 500 company." "I will be living in New York City, running my own data analytics startup." "I will be traveling around Canada in an RV while being an engineer for a fully distributed company."
- Reflect on your achievements in your current role that you've found most rewarding in the last twelve months. What are they, and how do they support your motivation for your future vision?
- What excites you most about your current role? How can that excitement be projected into your vision for the future?
- Do you see yourself managing people in the future? If so, what do you like most about it? What sort of people do you want to manage? Engineers or senior vice presidents? Venture capitalists?
- What are your formal qualifications? Do you use them in your current role and would you like to in the future? Do you want to get any more to make your future more compelling?

Take your time on this. Maybe let it sit for a while before you move on to the next section. Let your subconscious brain mull it over. Make sure that you're really happy with it. Here's a short example to get you thinking.

> **My Vision**
>
> *by: J. Bloggs*
>
> In ten years, I want to be the CTO of a technology company that I have co-founded. We will have a clear mission that does good for humanity and the planet. We favor purpose over profit. We will be a remote company, and I will be living with my family close to nature in a bright, spacious house. My partner and I will work from home, allowing us to spend close, quality time with our family.
>
> Over the last twelve months, I've experienced being a manager for the first time. I've helped junior staff tackle impostor syndrome and level up their skills, and I've been able to assume the role of a coach to my senior staff. I love helping people solve problems and seeing them grow as a result. It supports my future vision of being a CTO because I want to do this for an entire company.
>
> I'm most excited about the new product our team is launching later this year. I've never been responsible for shipping something so big, and I look forward to celebrating it with the team. I can't wait to do this on a bigger scale.
>
> In the future, I want to be managing people like me, showing them that the career path is rewarding and highly valued both by me and their staff. I want to help them help a whole department—located all over the world—succeed.
>
> My formal qualifications are few: I dropped out of college! However, I've shown that formal qualifications don't matter so much and I shouldn't worry about them. My career has moved the most when I have focused, been confident, and have applied myself. That's the same attitude I need to take into the future, because if I do, I know that I can achieve what I want.

Give it a go. Dream big, but don't worry about execution. Trying to stick to a ten-year plan inflexibly will make you unhappy because life is always changing. But having something to aim for *will* make you happy if you mindfully enjoy the journey rather than obsessing over the destination.

Get More Input!

If you want to make your vision even better, then seek input from others.

- Send your vision statement to your partner, a family member, a bunch of your friends, or even your colleagues or manager. Ask them for feedback. What do they think about it? Do they think it's realistic or overambitious? Could it even be under-ambitious?

Get More Input!

- Compile the social media profiles of five people that you think inhabit the future you desire. How do they present themselves? What are they working on right now? What do they post and talk about? Who do they interact with?

Your Plan

Now that you have your vision for the future in place, it's time to consider what your plan for the next year is. You'll do this by working out what you need to do in the short term to propel yourself toward your future vision, and you'll then create an actionable *skills backlog* to work through.

Writing Your Plan Statement

Firstly, reread your vision statement. Take it all in. With a pen and paper, jot down a few notes while you think about the following questions:

What will you have to do to reach your vision in the next

- Five years?
- Three years?
- Two years?
- One year?

Refer back to your vision at each stage and carefully think about the skills, opportunities, training, networking, and learning that you'll have to do. Where do you need to be at the end of the year to get there? Perhaps you just need to get your head down and become comfortable and confident in your new role. Perhaps you need to make the jump and get a job at a startup. Maybe you need to connect with more people in your local tech community. Either way, get it written down, and try to specify concrete actions that you need to take that you can provably demonstrate that you've achieved.

Here's an example.

My Plan

by: J. Bloggs

This year, I need to excel in my current role. I've been fortunate to have had the opportunity to be a manager, and if I want to be a CTO in the future, I need to learn the basics, implement them, and move along my path to mastering them.

I know that one of the areas that I often struggle with is knowing how to delegate effectively. I need to put that into practice as much as possible to increase the output of my team.

Additionally, I need to become confident at having challenging communications. I can do this by continually giving praise and critique whenever it's appropriate to do so.

I feel like I don't know many people doing this role in the local community, so I need to put some effort into connecting with them, both online and in person. I can do this by reaching out to them on social media, finding community groups online, and attending some local meetups.

I'll know that I've succeeded this year if I have a positive performance review from my manager and also supportive peer feedback from my team and peers.

Rating Your Competencies

Next, take a look at your career tracks that you defined in *Dual Ladders*, or use the career tracks from another company that you can find online. Find the role that you do and then take a note of the level that you currently rate yourself against each of the competencies.

For example:

Competency	Current Level
Technical knowledge	L3
Mentorship	L2
Influence	L2
Consensus building	L3
Communication	L2
Output	L2

Once you've done this, think about how working on your plan will advance your skills in each of these competencies. You'll be using this information to create your skills backlog.

Get More Input!

If you want to make your plan even better, then seek input from others.

- Send your plan statement to your manager, peers, or a trusted friend. What do they think about it? Does it feel correct and right-sized, or could it be improved?
- Ask for peer feedback on where you rank on each of the competencies. What do others think compared to your self-judgment? Why do you think that is? Do you need to change your rankings?

Creating Your Skills Backlog

With your plan and your rankings in hand, it's time to create your skills backlog for the next year. You're going to take the actionable items from your plan, map them against the competencies that they contribute toward, and record them somewhere that you can regularly update with your progress. Where exactly you keep them is up to you. A free solution is to use a tool like Trello.[3] This will allow you to make comments on the items as they progress through your board.

Your backlog could look something like this:

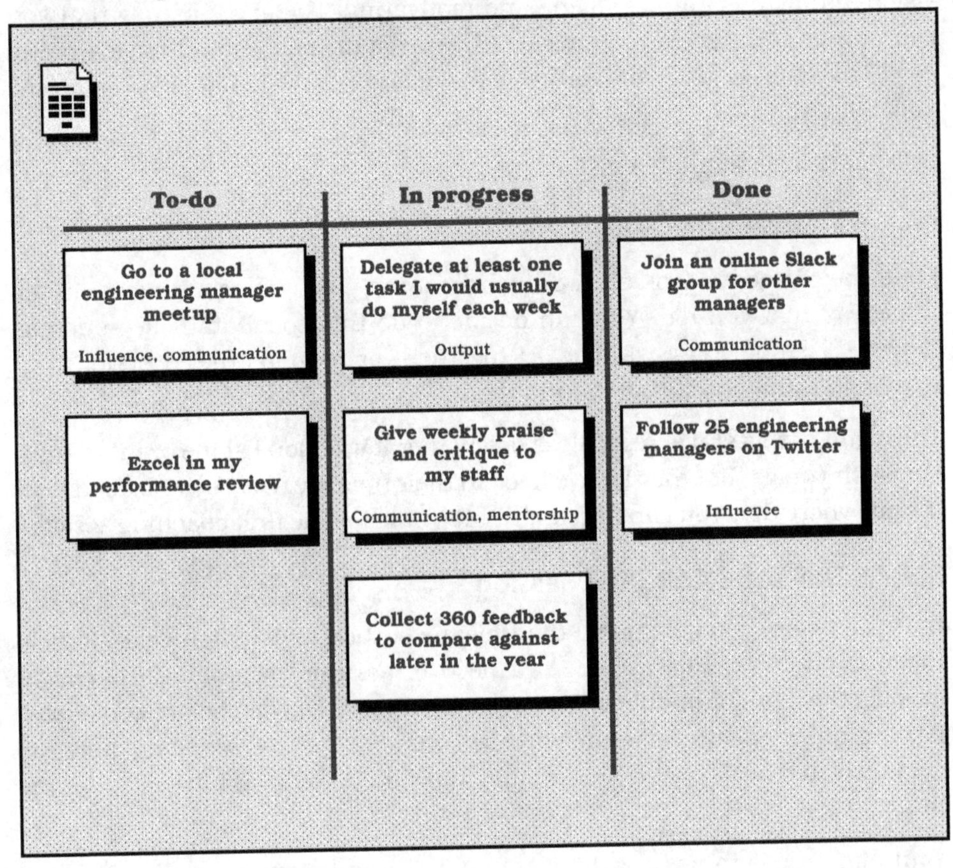

Refer to your skills backlog regularly from now on. Check to see if you're making progress. If not, why? Your vision of the future needs you! It's useful to refer back to the items on your backlog with your manager quarterly so

3. https://trello.com/

you can get their input. It may be the case that you can bring in new skills to work on, and you can prune away those that you've completed.

Well done! That brings us to the end of the exercise. You started by dreaming up a ten-year vision, and then you narrowed it down to a plan for the year. Then you derived key skills to work on, which you can revisit as often as you like. Neat, huh?

Performing the Exercise with Your Staff

I'm sure you'll have seen that the exercise above in which you define your vision and plan is both engaging and motivating. The good news is that you can now go through this exercise with your staff. You should have a pretty good idea how to do it as you've just done it yourself, but here are some additional tips.

Let's start with defining a vision.

Step 1: Ten-Years Exercise

First, they'll need to look back over the last ten years using a timeline in the same way that you did. You can decide to do this collaboratively—such as drawing it all out on a whiteboard together—or perhaps they'd prefer to go away and think about it. It's up to them.

The result of this stage, regardless of whether it was done alone or in partnership with you, is that they'll have filled in their timeline. Get them to document it somewhere that you can find it later. It's time for the first coaching session.

Step 2: Ten-Years Coaching Session

In this session, you'll be using the timeline of the previous ten years to help them dream about the future. You need to assume the role of the coach, keeping the thought bubble over their head as we've learned before and asking leading questions to help them realize that ten years is a long time, as demonstrated by the *past*. This should allow them to be ambitious about the *future*.

Your aim here is to use coaching techniques and language to allow them to think *creatively* and to think *big*—that is, beyond the next few promotions and pay raises. You can borrow the approach that we took when we went through the exercise together in *What About the Next Ten Years?*

As you go through those questions, try to be provocative to get them to think at the right scale, such as "what skills do you think that the world is in danger of overlooking?" It's common for people to understate themselves, especially if they're introverted. So use your intelligence to listen *beyond* the words that they're telling you. Probe deeper to find the true meaning. "I want to work in Product" may *actually* be "I want to be the chief product officer at my own successful startup."

Encourage them to share the truth with you. Ask questions. As you're moving through the exercise, make notes together. Regularly summarize what they're saying back to them to achieve consensus: "So what I hear is.... Is that a fair summary?"

You should both find this session exciting: the content should be meaningful and will help you learn a lot about what your direct report wants to achieve both in their professional and personal life. Once you're done, it's time for them to do their homework: they'll need to write their vision statement, similar to the example in *My Vision*. Review that section again for a reminder of how you did it. They should send it to you to review after they're done.

Step 3: Reviewing Their Vision

Next, you'll book another session and sit down with their vision statement. Discuss what you thought about it, and also how they feel now that some time has passed since they wrote it. Does it resonate? Does it feel authentic? Is it actually a vision, or just a laundry list of things to do? Are there any parts of it that you want to dig into deeper to increase your understanding or to make them more compelling?

At the end of this session, in addition to prompting any further revisions to their statement, you can help them seek additional validation in the same way that you did. Get them to research online to find the profiles of other people who have successfully achieved a similar position. Get them to research how they got there, what content they produce and share, and who they talk to. Get them to immerse themselves in a future that they can manifest. Additionally, you can prompt them to email their vision statement to five friends to get more feedback from people that are close to them.

Step 4: Input, Scoring, and Homework

With the career tracks in hand, get them to rate the level at which they reside for each competency. Now it's peer feedback time. With their collaboration, choose a handful of colleagues and ask for them to perform the exact same competency-mapping exercise. Collect the answers. Once that's done, book in another session.

In preparation for that session, they'll need to think about the next six months by iterating through the next five, three, and two years, like you did in *Writing Your Plan Statement*. They should bring their first attempt at their plan statement for you to review.

Step 5: One-Year Coaching Session

To begin with, review the competency rankings that they did for themselves and also those that came back as part of the feedback process. If there's a large gap between where they think they are and where others see them, then do use your tact to deliver the message in a radically candid way.

Collaboratively read through their plan statement. Pick out the concrete actions that they'll need to focus on and map them where possible to competencies on the career ladder. These will become the items that they'll be placing in their skills backlog. Ensure that they are tangible and measurable tasks to work from.

Their homework is to continually review their skills backlog and to ensure that they're doing work that makes progress in this area. Your homework will be to make sure that you're revisiting their skills backlog at least every three months, and refining their plan and backlog items every six.

Your staff should now have all of the tools they need to propel their own career forward with your support. Not bad for a few coaching sessions.

That's All Folks!

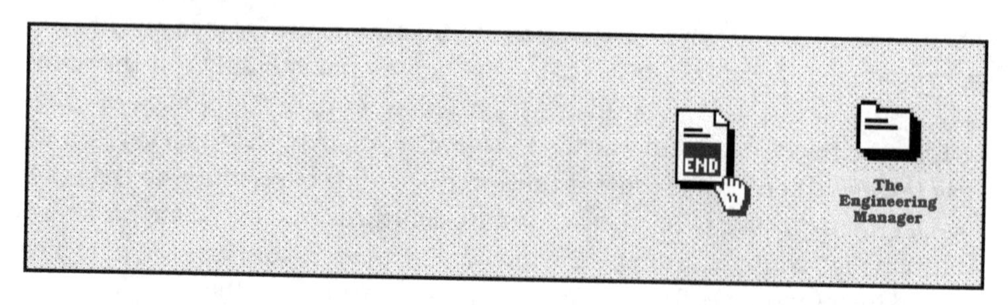

Wow, is this really the end of our time together? We're at the end of the chapter and also at the end of the entire book. But before I get emotional, let's quickly review what we learned in this chapter:

- We saw that having a compelling *vision* for the future, backed up by a short-term actionable *plan*, is a key ingredient for the success and happiness of you and your staff.

- We went through a multistage *exercise* in which you were able to create these for yourself, culminating in a *skills backlog* that you can continually work on.
- Then, we looked at how you can provide this same exercise for your staff through a series of *coaching sessions*. You can be more than just their manager: you can be their career coach too.

In the *Introduction*, we said that the technology industry was facing a skills crisis—not because we don't know how to write code, but because we don't know how to manage humans. I hope that now you've been through this book, you feel like you're on your way to being a confident, productive, effective manager, ready to make a real difference in the lives of your staff. What's more is that you're in a great position to help others who may want to take the same path as you in the future. One by one, we can make the technology industry better. We really can.

Now it truly is time to say goodbye. I've enjoyed sharing this journey with you. I hope that the skills you've learned in this book take you far in your career and that you're able to continually build upon them until you're able to make whatever you dreamed up in your vision a reality.

If you pick up these pages in the future, we'll be in touch again then. However, until next time, let me say this—it's been a pleasure. Best of luck. And remember, *you've got this*.

Bibliography

[Ash19] Jason Allen Ashlock. Why Turnover is a Good Thing for Your Company. *Inc.*. 2019, September.

[Ber13] Scott Berkun. *The Year Without Pants: WordPress.com and the Future of Work.* John Wiley & Sons, New York, NY, 2013.

[Boo18] Michael Booz. These 3 Industries Have the Highest Talent Turnover Rates. *LinkedIn.* 2018, March.

[Bro95] Frederick P. Brooks Jr. *The Mythical Man-Month: Essays on Software Engineering.* Addison-Wesley, Boston, MA, Anniversary, 1995.

[Bue10] Dan Buettner. *Blue Zones: Lessons for Living Longer from the People Who've Lived the Longest.* National Geographic Society, Washington, D.C., 2010.

[Cam19] University of Cambridge. Mentoring Agreement. *University of Cambridge, Personal and Professional Development.* 2019, September.

[Chr16] Clayton M. Christensen. *The Innovator's Dilemma (Management of Innovation and Change).* Harvard University Press, Boston, MA, 2016.

[CI78] Pauline Rose Clance and Suzanne Imes. The Imposter Phenomenon in High Achieving Women: Dynamics and Therapeutic Intervention. *Psychotherapy Theory, Research and Practice.* 15[3], 1978.

[DGS13] Cristina Díaz-García, Angela González-Moreno, and Francisco Jose Sáez-Martínez. Gender diversity within research and development teams: Its impact on radicalness of innovation. *Innovation.* 15[2], 2013.

[DLS06] Esther Derby, Diana Larsen, and Ken Schwaber. *Agile Retrospectives: Making Good Teams Great.* The Pragmatic Bookshelf, Raleigh, NC, 2006.

[Dow14] Myles Downey. *Effective Modern Coaching.* LID Publishing, London, UK, 2014.

[FH13] Jason Fried and David Heinemeier Hansson. *Remote: Office Not Required.* Penguin, New York, NY, 2013.

[Fou17] Camille Fournier. *The Manager's Path: A Guide for Tech Leaders Navigating Growth and Change.* O'Reilly & Associates, Inc., Sebastopol, CA, 2017.

[GFK11] Danielle Gaucher, Justin Friesen, and Aaron, C. Kay. Evidence that gendered wording in job advertisements exists and sustains gender inequality. *Journal of Personality and Social Psychology.* 101[1], 2011.

[Git19] Gitlab. Gitlab Handbook. *Gitlab.* 2019, December.

[Gra09] Paul Graham. Maker's Schedule, Manager's Schedule. *Essays by Paul Graham.* 2009, July.

[Gro95] Andrew S. Grove. *High Output Management.* Vintage Books, New York, NY, 1995.

[Gro98] Andrew S. Grove. *Only The Paranoid Survive.* HarperCollins Publishers, New York, NY, 1998.

[HCY14] Reid Hoffman, Ben Casnocha, and Chris Yeh. *The Alliance: Managing Talent in the Networked Age.* Harvard University Press, Boston, MA, 2014.

[Hor19] Ben Horowitz. *What You Do Is Who You Are: How to Create Your Business Culture.* William Collins, London, UK, 2019.

[HT00] Andrew Hunt and David Thomas. *The Pragmatic Programmer: From Journeyman to Master.* Addison-Wesley, Boston, MA, 2000.

[Hun08] Andy Hunt. *Pragmatic Thinking and Learning: Refactor Your Wetware.* The Pragmatic Bookshelf, Raleigh, NC, 2008.

[Irv09] William B Irvine. *A Guide to the Good Life: The Ancient Art of Stoic Joy.* OUP USA, New York, USA, 2009.

[Kab13] Jon Kabat-Zinn. *Full Catastrophe Living, Revised Edition: How to cope with stress, pain and illness using mindfulness meditation.* Piatkus, London, UK, 2013.

[Kab16] Jon Kabat-Zinn. *Wherever You Go, There You Are: Mindfulness meditation for everyday life.* Piatkus, London, UK, 2016.

[Kah12] Daniel Kahneman. *Thinking, Fast and Slow.* Penguin, New York, NY, 2012.

[KD99] Justin Kruger and David Dunning. Unskilled and Unaware of It: How Difficulties in Recognizing One's Own Incompetence Lead to Inflated Self-

Assessments. *Journal of Personality and Social Psychology*. 77[6]:1121–1134, 1999.

[Kut13] Joe Kutner. *The Healthy Programmer: Get Fit, Feel Better, and Keep Coding*. The Pragmatic Bookshelf, Raleigh, NC, 2013.

[KZK16] Jake Knapp, John Zeratsky, and Braden Kowitz. *Sprint: How To Solve Big Problems and Test New Ideas in Just Five Days*. Sentinel, New York, USA, 2016.

[Lal14] Frederic Laloux. *Reinventing Organizations: A Guide to Creating Organizations Inspired by the Next Stage in Human Consciousness*. Nelson Parker, Brussels, Belgium, 2014.

[Mur18] Christopher J. L. Murray. Global, regional, and national incidence, prevalence, and years lived with disability for 354 diseases and injuries for 195 countries and territories, 1990–2017: a systematic analysis for the Global Burden of Disease Study 2017. *The Lancet*. 392, 2018.

[MWC19] Sarah Myers West, Meredith Whittaker, and Kate Crawford. Discriminating Systems: Gender, Race and Power in AI. *AI Now Institute whitepaper*. 2019.

[Nat18] National Academies of Sciences, Engineering and Medicine. *Sexual Harassment of Women: Climate, Culture, and Consequences in Academic Sciences, Engineering, and Medicine*. National Academies Press (US), Washington (DC), 2018.

[OS14] Marily Oppezzo and Daniel L. Schwartz. Give your ideas some legs: The positive effect of walking on creative thinking. *Journal of Experimental Psychology: Learning, Memory, and Cognition*. 2014, July.

[Pin09] Daniel H. Pink. *Drive: The Surprising Truth About What Motivates Us*. Riverhead Books, New York, NY, 2009.

[Ray01] Eric S. Raymond. *The Cathedral and The Bazaar*. O'Reilly & Associates, Inc., Sebastopol, CA, 2001.

[Rog12] Simon Rogers. Two-body problem (career). *The Guardian*. 2012, August.

[Rum11] Donald Rumsfeld. *Known And Unknown*. Bantam Press, London, UK, 2011.

[San13] Sheryl Sandberg. *Lean In: Women, Work, and the Will to Lead*. Random House, New York, NY, 2013.

[Sco17] Kim Scott. *Radical Candor: Be a Kickass Boss Without Losing Your Humanity*. St. Martin's Press, New York, NY, 2017.

[SFKI16] Bryan Sisk, Richard Frankel, Erik Kodish, and J Harry Isaacson. The Truth about Truth-Telling in American Medicine: A Brief History. *The Permanente Journal*. 20[3], 2016.

[Sre05] Sampath Venkata Sreekanth. Mindfulness based stress reduction in medicine. *BMJ.* 331[40], 2005.

[SSAR14] Roderick I. Swaab, Michael Schaerer, Eric M. Anicich, Richard Ronay, and Adam D. Galinsky. The Too-Much-Talent Effect: Team Interdependence Determines When More Talent Is Too Much or Not Enough. *Psychological Science.* 25[8], 2014.

[Wal17] Matthew Walker. *Why We Sleep: The New Science of Sleep and Dreams.* Penguin, New York, NY, 2017.

Index

Thank you!

How did you enjoy this book? Please let us know. Take a moment and email us at support@pragprog.com with your feedback. Tell us your story and you could win free ebooks. Please use the subject line "Book Feedback."

Ready for your next great Pragmatic Bookshelf book? Come on over to https://pragprog.com and use the coupon code BUYANOTHER2020 to save 30% on your next ebook.

Void where prohibited, restricted, or otherwise unwelcome. Do not use ebooks near water. If rash persists, see a doctor. Doesn't apply to *The Pragmatic Programmer* ebook because it's older than the Pragmatic Bookshelf itself. Side effects may include increased knowledge and skill, increased marketability, and deep satisfaction. Increase dosage regularly.

And thank you for your continued support,

Andy Hunt, Publisher

Build Websites with Hugo

Rediscover how fun web development can be with Hugo, the static site generator and web framework that lets you build content sites quickly, using the skills you already have. Design layouts with HTML and share common components across pages. Create Markdown templates that let you create new content quickly. Consume and generate JSON, enhance layouts with logic, and generate a site that works on any platform with no runtime dependencies or database. Hugo gives you everything you need to build your next content site and have fun doing it.

Brian P. Hogan
(154 pages) ISBN: 9781680507263. $26.95
https://pragprog.com/book/bhhugo

Practical Microservices

MVC and CRUD make software easier to write, but harder to change. Microservice-based architectures can help even the smallest of projects remain agile in the long term, but most tutorials meander in theory or completely miss the point of what it means to be microservice based. Roll up your sleeves with real projects and learn the most important concepts of evented architectures. You'll have your own deployable, testable project and a direction for where to go next.

Ethan Garofolo
(290 pages) ISBN: 9781680506457. $45.95
https://pragprog.com/book/egmicro

Real-Time Phoenix

Give users the real-time experience they expect, by using Elixir and Phoenix Channels to build applications that instantly react to changes and reflect the application's true state. Learn how Elixir and Phoenix make it easy and enjoyable to create real-time applications that scale to a large number of users. Apply system design and development best practices to create applications that are easy to maintain. Gain confidence by learning how to break your applications before your users do. Deploy applications with minimized resource use and maximized performance.

Stephen Bussey
(326 pages) ISBN: 9781680507195. $45.95
https://pragprog.com/book/sbsockets

Programming Machine Learning

You've decided to tackle machine learning — because you're job hunting, embarking on a new project, or just think self-driving cars are cool. But where to start? It's easy to be intimidated, even as a software developer. The good news is that it doesn't have to be that hard. Master machine learning by writing code one line at a time, from simple learning programs all the way to a true deep learning system. Tackle the hard topics by breaking them down so they're easier to understand, and build your confidence by getting your hands dirty.

Paolo Perrotta
(340 pages) ISBN: 9781680506600. $47.95
https://pragprog.com/book/pplearn

Competing with Unicorns

Today's tech unicorns develop software differently. They've developed a way of working that lets them scale like an enterprise while working like a startup. These techniques can be learned. This book takes you behind the scenes and shows you how companies like Google, Facebook, and Spotify do it. Leverage their insights, so your teams can work better together, ship higher-quality product faster, innovate more quickly, and compete with the unicorns.

Jonathan Rasmusson
(138 pages) ISBN: 9781680507232. $26.95
https://pragprog.com/book/jragile

Programming Flutter

Develop your next app with Flutter and deliver native look, feel, and performance on both iOS and Android from a single code base. Bring along your favorite libraries and existing code from Java, Kotlin, Objective-C, and Swift, so you don't have to start over from scratch. Write your next app in one language, and build it for both Android and iOS. Deliver the native look, feel, and performance you and your users expect from an app written with each platform's own tools and languages. Deliver apps fast, doing half the work you were doing before and exploiting powerful new features to speed up development. Write once, run anywhere.

Carmine Zaccagnino
(368 pages) ISBN: 9781680506952. $47.95
https://pragprog.com/book/czflutr

Agile Web Development with Rails 6

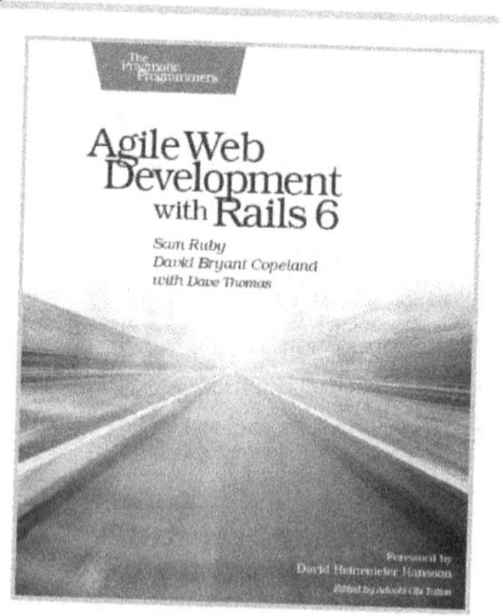

Learn Rails the way the Rails core team recommends it, along with the tens of thousands of developers who have used this broad, far-reaching tutorial and reference. If you're new to Rails, you'll get step-by-step guidance. If you're an experienced developer, get the comprehensive, insider information you need for the latest version of Ruby on Rails. The new edition of this award-winning classic is completely updated for Rails 6 and Ruby 2.6, with information on processing email with Action Mailbox and managing rich text with Action Text.

Sam Ruby and David Bryant Copeland
(494 pages) ISBN: 9781680506709. $57.95
https://pragprog.com/book/rails6

Modern Systems Programming with Scala Native

Access the power of bare-metal systems programming with Scala Native, an ahead-of-time Scala compiler. Without the baggage of legacy frameworks and virtual machines, Scala Native lets you re-imagine how your programs interact with your operating system. Compile Scala code down to native machine instructions; seamlessly invoke operating system APIs for low-level networking and IO; control pointers, arrays, and other memory management techniques for extreme performance; and enjoy instant start-up times. Skip the JVM and improve your code performance by getting close to the metal.

Richard Whaling
(260 pages) ISBN: 9781680506228. $45.95
https://pragprog.com/book/rwscala

Fixing Your Scrum

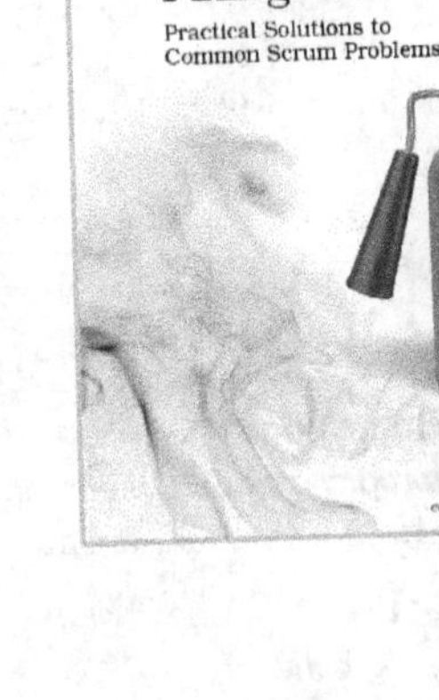

Broken Scrum practices limit your organization's ability to take full advantage of the agility Scrum should bring: The development team isn't cross-functional or self-organizing, the product owner doesn't get value for their investment, and stakeholders and customers are left wondering when something—anything—will get delivered. Learn how experienced Scrum masters balance the demands of these three levels of servant leadership, while removing organizational impediments and helping Scrum teams deliver real-world value. Discover how to visualize your work, resolve impediments, and empower your teams to self-organize and deliver using advanced coaching and facilitation techniques that honor and support the Scrum values and agile principles.

Ryan Ripley and Todd Miller
(240 pages) ISBN: 9781680506976. $45.95
https://pragprog.com/book/rrscrum

Software Estimation Without Guessing

Developers hate estimation, and most managers fear disappointment with the results, but there is hope for both. You'll have to give up some widely held misconceptions: let go of the notion that "an estimate is an estimate," and estimate for your particular need. Realize that estimates have a limited shelf-life, and re-estimate frequently as needed. When reality differs from your estimate, don't lament; mine that disappointment for the gold that can be the longer-term jackpot. We'll show you how.

George Dinwiddie
(246 pages) ISBN: 9781680506983. $29.95
https://pragprog.com/book/gdestimate

The Pragmatic Bookshelf

The Pragmatic Bookshelf features books written by professional developers for professional developers. The titles continue the well-known Pragmatic Programmer style and continue to garner awards and rave reviews. As development gets more and more difficult, the Pragmatic Programmers will be there with more titles and products to help you stay on top of your game.

Visit Us Online

This Book's Home Page

https://pragprog.com/book/jsengman

Source code from this book, errata, and other resources. Come give us feedback, too!

Keep Up to Date

https://pragprog.com

Join our announcement mailing list (low volume) or follow us on twitter @pragprog for new titles, sales, coupons, hot tips, and more.

New and Noteworthy

https://pragprog.com/news

Check out the latest pragmatic developments, new titles and other offerings.

Save on the ebook

Save on the ebook versions of this title. Owning the paper version of this book entitles you to purchase the electronic versions at a terrific discount.

PDFs are great for carrying around on your laptop—they are hyperlinked, have color, and are fully searchable. Most titles are also available for the iPhone and iPod touch, Amazon Kindle, and other popular e-book readers.

Send a copy of your receipt to support@pragprog.com and we'll provide you with a discount coupon.

Contact Us

Online Orders:	*https://pragprog.com/catalog*
Customer Service:	*support@pragprog.com*
International Rights:	*translations@pragprog.com*
Academic Use:	*academic@pragprog.com*
Write for Us:	*http://write-for-us.pragprog.com*
Or Call:	+1 800-699-7764

CPSIA information can be obtained
at www.ICGtesting.com
Printed in the USA
JSHW031743050122
21823JS00004B/52